Industrial Sociology

A Comprehensive Approach

Er. Osama Lari
Lecturer
Mechanical Engineering Department
Integral University
Lucknow

CBS

CBS Publishers & Distributors Pvt Ltd

New Delhi • Bengaluru • Chennai • Kochi • Pune
Hyderabad • Kolkata • Mumbai • Nagpur • Patna

Disclaimer

Science and technology are constantly changing fields. New research and experience broaden the scope of information and knowledge. The author has tried his best in giving information available to him while preparing the material for this book. Although, all efforts have been made to ensure optimum accuracy of the material, yet it is quite possible some errors might have been left uncorrected. The publisher, the printer and the author will not be held responsible for any inadvertent errors, omissions or inaccuracies.

Industrial Sociology
A Comprehensive Approach

ISBN: 978-81-239-2291-1

CBS First Edition: 2013

Published by Satish Kumar Jain for
CBS Publishers & Distributors Pvt Ltd
4819/XI Prahlad Street, 24 Ansari Road, Daryaganj, New Delhi 110 002, India.
Ph: 23289259, 23266861, 23266867 Fax: 011-23243014 Website: www.cbspd.com
e-mail: delhi@cbspd.com; cbspubs@airtelmail.in

Corporate Office: 204 FIE, Industrial Area, Patparganj, Delhi 110 092
Ph: 4934 4934 Fax: 4934 4935 e-mail: publishing@cbspd.com; publicity@cbspd.com

Branches

- **Bengaluru:** Seema House 2975, 17th Cross, K.R. Road, Banasankari 2nd Stage, Bengaluru 560 070, Karnataka
  Ph: +91-80-26771678/79 Fax: +91-80-26771680 e-mail: bangalore@cbspd.com
- **Chennai:** 20, West Park Road, Shenoy Nagar, Chennai 600 030, Tamil Nadu
  Ph: +91-44-26260666, 26208620 Fax: +91-44-42032115 e-mail: chennai@cbspd.com
- **Kochi:** 36/14 Kalluvilakam, Lissie Hospital Road, Kochi 682 018, Kerala
  Ph: +91-484-4059061-65 Fax: +91-484-4059065 e-mail: kochi@cbspd.com
- **Pune:** Bhuruk Prestige, Sr. No. 52/12/2+1+3/2 Narhe, Haveli (Near Katraj-Dehu Road Bypass), Pune 411 041, Maharashtra
  Ph: +91-20-64704058, 64704059, 32342277 Fax: +91-20-24300160 e-mail: pune@cbspd.com

Representatives

- **Hyderabad** 0-9885175004
- **Kolkata** 0-9831437309, 0-9051152362
- **Mumbai** 0-9833017933
- **Nagpur** 0-9021734563
- **Patna** 0-9334159340

Printed at India Binding House Noida

In the Name Of "Allah" the Beneficent and Merciful

This Book is Dedicated to my Father Mr. Gufran Ahmad Lari , My mother Ruhi Tabassum, My Sister Tahseen Fatima , My Brothers Anas Lari . Asad Lari . Zeeshan Lari and to My uncle Late Mr. Rizwan Ahmad Lari (Gauri)

Er. Osama Lari

Syllabus

Industrial Sociology
(EHU-302/EHU-402)
U.P. Technical University

Unit-I

Industrial Sociology : Nature and Scope of Industrial Sociology-Development of Industrial Sociology.

Unit-II

Rise and Development aof Industry : Early Industrialism – Types of Productive Systems – The Manorial or Feudal system – The guild system – The domestic or putting-out system – and the factory system – Characteristics of the factory system – causes and Consequences of industrialization.

Unit-III

Industrialization in India. Industrial Poling Resolutions – 1956.

Unit-IV

Contemporary Issues : Grievances and Grievance handling Procedure.

Industrial Disputes : courses, strikes & lockouts, Industrial Relations Machinery Bi-partite & Tri-partite Agreement, Labour courts & Industrial Tribunals, Code of Discipline, Standing order.

Preface

This text Book on "Industrial Sociology" has been designed to fulfill the needs of Second year students of Engineering discipline of U.P. Technical University, Integral University and many other Technical Universities in India.

This book presents succinct coverage of theory, definitions, quotes and concerned references. It has plenty of clear cut diagrams, pictures and cartoons to make the therein mentioned principles easily comprehensible and attractive.

The main objective of this book is to provide a bunch of course material based on prescribed syllabus and to develop the ability to analyze any term in a simple and logical manner. It is hoped that the book in its present form shall be preferred and appreciated by the readers due to its well researched and authentic material. All the suggestions regarding this book from the readers for any improvement or correction or any study material addition are welcome.

Er. Osama Lari

Acknowledgement

I would like to express my heartfelt gratitude to all, those who helped and encouraged me in writing this book. First of all, I would like to express my thanks to Visionary Maestro Honorable Vice-Chancellor of Integral University, Prof. S.W. Akhtar and Pro-Vice Chancellor Prof. S.M. Iqbal and Registrar Dr. I.A. Khan for their kind blessings and encouragement.

I would like to express my warm gratitude to Dr. M. I. Khan, Dean and HOD Mechanical Engg. Department, Er. P.K. Bharti, Er. K.M. Moeed, Er. SAH Jafri, Er. Shahnawaz Alam, Er. Neeraj Gupta, Er. Mohd Shadab, Er. Anas, Er. Ashar Ali Minai, Er. Abdul Ahed khan, Er. Yaqoob Ali Ansari, Er. Serajul Haque, Er. Anwar Ahmad Sheikh, Er. Ather Hussain, Er. Saquib Ali, Er. Mahmood Alam, Er. Mahmood Khan, Er. Suhail Ahmad, Er. Khalid, Ahmad Safuaan, Nesar Ahmad and Mohd Amir of Integral University and last but not the least to my well wishers Shahjan Asif, Sana Tabassum, Shakilurahman Lari, Irshad Ahmad Lari, Dr. Javed Ahmad Lari, Sarwat Mumtaz, Zaki Ahmad Lari, Mrs. Zakia Rahman, Miss Neha, Miss Bushra Ahmad, Fazeel Raees Khan, Danish Kamal Lari, Afzal, Nazeeb Alam and Md. Ayaz Warsi for guiding and helping me in the preparation and completion of this book.

I would also thank Mr. Suneel Gomber and Mr. Sushant Gomber for their marvelous effort to print this book in record time with an excellent presentation style.

Er. Osama Lari

Special Thanks to....

Er. Gauhar Raza
Bhabha Atomic Research Center, Mumbai

Er. Mohd. Anas Siddiqui
Renautlt Nissan, Chennai

Er. Praveen Singh
UPRVNL, Obra

Er. Misbahul Khan
EDS, Solutions, Delhi

Contents

Chapter 1
Industrial Sociology

1.1 Introduction

The industrialisation and urbanisation had a profound influence on the emerging sociological disciplines. Today's Economy is changing from an economy of small family farms to one of giant corporations as every one comes to a single platform in order to earn their livelihood. People are working in the field or may work in an office as salaried personnel or may work as a owner in any industry or organisation where many personnel's work together. All the above platforms require discipline i.e. social discipline in order to maintain the status and dignity in society or amongst industries.

1.2 Quotes on Sociology

Here given below are some famous quotes laid down by some famous sociologists.

(a) "We may begin by noting that sociologists increasingly fail to ask what is worth knowing and what is not (Moore 1998). Instead they either plunge headlong into highly specialized and methodologically sophisticated investigations of matters of moderate significance or pursue zealously and with an apparently satisfying sense of resentment the political-ideological issues bequeathed to us by the 1960s." –Paul Hollander

(b) "In fact, for various reasons, academic sociology has largely been transformed from the applied science envisioned by the founder's into what is essentially a literary pursuit. The journal article especially has become the essential product of one's Labors, and with good reason. For it is the article, and the article alone, that, if published in the right place, can secure a good job, win tenure, ensure promotion, and --if widely cited--gain one a sort of immortality (or at least 15 minutes of fame) –Jay Weinstein

1.3 Definition of Sociology

Sociology is the scientific study of human social behavior, interaction between humans, social institutions and social organisation/structure. In other words it is a scientific study of society and human behavior. In this, society plays a role of common platform on which various theories of society itself and behavior if human can be easily understood.

The word sociology was derived from a Latin word: 'Socius' means "companion" and a suffix---ology means "the study of" and Greek *logos*, "word", "knowledge". It was first coined in 1780 by the French essayist Emmanuel Joseph Sieyes (1748–1836) in an unpublished manuscript. Sociology was later developed by the philosopher, Auguste Comte (1798–1857), in 1838. Comte had earlier used the term "social physics", but that had subsequently been appropriated by others, most notably the Belgian statistician Adolphe Quetelet. Comte endeavoured to unify history, psychology and economics through the scientific understanding of the social realm. Writing shortly after the malaise of the French Revolution, he proposed that social ills could be remedied through sociological positivism, an epistemological approach outlined in *The Course in Positive Philosophy* [1830–1842] and *A General View of Positivism* (1844). Comte believed a positivist stage would mark the final era, after conjectural theological and metaphysical phases, in the progression of human understanding.

Sociology examines the relations between all aspects of social life i.e. social facts and problems, in modern industrialized society, but with special emphasis on class, race and gender. Because of this broad scope, it is the most inclusive of the social sciences. In fact, if sociology is the "Systematic study of collective human behavior", it would be "The Social Science".

Perhaps unique in the social sciences, sociology is concerned with the character and dynamics of modern, industrial society like anthropology, sociology focuses on culture-ways of thinking, feeling and acting learned from others as members of a group.

1.4 History/Development of Sociology

In 1600's, Political arithmetic began in Great Britain, as demographic and census activity increased, data became available on the nature of social problems from 1815-1848, Public hygienists and medical reformers became increasingly concerned with the condition of the working of lower class in the urban industrial environment. Moral statisticians could use demographic and other census data to generate theories to explain population and social problem including crime, suicide and pauperism. As interest grew, there was increased demand for more detailed data sets and for scholars to interpret them. It its beginning, sociology was most concerned with solving immediate, highly visible urban problems.

Between 1830-1842, Augusta Comte's cours de Philosophie positive gave sociology a name and place among the academic disciplines. Comte, long interested in the history of human society, saw sociology as the master science of human behavior which would discover the laws of human behavior, (the laws of human society) just as natural science discovered the laws of nature. There was considerable emphasis on morality and moral progress. Scientific thought would guide social planning. Sociology would organise and systematize all human knowledge. Knowledge gained from Sociological research would render government and human affairs more efficient. Society was seen as a set of functional relationships. One of the Comte's most important colleagues was Henride Saint-Simon who was one of the founders of socialism.

August Comte :

Weathering a Storm of Change

(1798-1857) grew up in the wake of the French Revolution and industrial revolution which brought a sweeping transformation to his country. He believed, people would be able to build for themselves a better future. He divided his new discipline into social statics i.e. how society is held together and how society changes (social dynamics). From the Greek and Latin words meaning "the study of society," Comte came to describe his work as sociology.

In the last part of the 19th century, sociology was increasingly recognized as an academic discipline in Europe-Emile Durkheim became highly visible and promoted sociological research and focused on "human behavior as structure and process", discuss, control and power in human affairs," and Society as a "Set of functional relationships".

Emile Durkheim:

Unmasking the Power of Society

He propounded that one can understand people, not by looking at individuals, but only by examining their society. Durkheim's investigation of suicide, detailed in Chapter 1 of society's power to shape human behavior i.e. people's place within the social system as women or men, rich or poor, Catholic, Jew, or Protestant-affects even this most personal act. His contributions to the understanding of crime figure prominently in Chapter 8 ("Deviance"). Durkheim also spent much of his life investigating religion, which he held to be a key foundation of social integration (see Chapter 18, "Religion"). Just as important, Durkheim is also one of the major architects of the structural functional paradigm, which we refer to in almost every chapter that follows.

During the same time, sociology emerged in the U.S. curiously, contemporary sociology is mostly American. The first U.S. sociology course was taught at Yale in 1875. By 1893, the first U.S. sociology department was established at the University of Chicago. This became the center of U.S. sociology until the 1940s. A year later in 1894, the first U.S. Sociology text book was published. The following year, the first issue of the American

Journal of Sociology was issued by the University of Chicago. By 1900, some large U.S. universities had sociology departments. In 1905, the American sociological society was established. Still, the discipline began with practical aspects of sociology such as criminology and industrial relations began to distance themselves from the academic sociology department. The two decades after World War II were a time of great expansion for sociology with more departments established and growing enrollments. In 1949, under sponsorship of UNESCO, the International sociological Association was established, with Louis wirth of Chicago University as its First president.

By the 1960's and 1970's, more than a dozen sociological journals and many minor journals were published because of this interest in the practical aspects of sociology, especially in social services. This expansion had allowed a small but increasing number of sociologists to pursue their full time researches and Degree's to the Sociologists were awarded by some Leading institutions like.

- University of California, Berkeley
- Harvard University
- University of Chicago
- Columbia University
- University of Michigan
- University of Wiscongin
- University of North Carolina, Chapel thill and many more

Sociology does not provide solutions to the present problems; it is only believe of Sociologists that with increasing the applications of Social Science to these problems, the goal of human welfare can be achieved. i.e. factors like corruption, crime, family break down, poverty and racial/ethnic minorities which are against the welfare of human's involved in any family or organisation can be removed.

1.5 Scope/Problem's oriented by Sociology

Sociologists employs various method of social quests in order to describe various social behaviours, relationships, changes in social values and reactions of personnel's during social value change. Sociologists are divided into two groups-Micro sociologists, who looks at particular interactions and Macro sociologists, who looks at the pattern of interactions, usually in larger groups. Norms or expectations or rules or models of conduct are of particular interest to Sociologists. These vary from group to group as does social structure which is the pattern of recurring interaction among individuals with different roles.

With the birth of Industrial age, Sociology has a long history of interest in solving the problems associated with urbanisation and industrialisation. Such problems include.

- Poverty

- Inequality
- Family breakdown
- Crime/deviant
- Racial/ethnic minorities

This problem orientation has led to a focus on the "Victims" of modern industrial society. Women, racial and ethnic groups have received considerable attention.

1.6 Application of Sociology

Sociology had developed strong linkage to the "welfare" State and to the professions that deal with Social problems: Social Work, Social administration, education and public health.

Various Sociologists are employed in different bodies of society at domestic or international level for following reasons:

- Gathering information for planning and process.
- Evaluation of progress in Society.
- Proposing different methods of change.
- Dealing with issues of population or budgets.
- Sociologists performs actions like social surveys, solves public relation problem, advices to various consumer and many more for industries.
- Main area of application of sociology in Academic institutions or orgánisation.
- Sociologists play a role of counselor while solving minority or racial problems.
- Help in understanding and predicting the human behavior in different situations.
- Also helps in understanding the influence of society on any human and what make them come together to live in a society.

1.7 Inter-disciplinary of Sociology

Sociology maintains the closest relationship to other disciplines that also helps in studying about various aspects of Society. Various fields like anthropology, economics, History, Psychology, Sociobiology and many more consists very close relation.

Sociology is much more similar to anthropology because both the disciplines seek to understand human society as a whole or we can say study of human differences like cultural or biological in the context of human nature. Anthropologists identify and compare behavior of a particular group against the full range of human behavior i.e. social facts which includes way of life, language and physical characteristics of people.

Similarly, Psychology deals with the study of an individual in any group or the relations of groups to one another. Psychology includes the study of the human mind, his social attitudes and opinions. Both Psychology and sociology shares a common interest of subfield social Psychology. Where sociology deals with the various behavior of human in state of any problem with other member's of group or family and Psychology deals with working state of human mind i.e. the way with which problem is being handled or action's taken in that state of problem or hard situation.

Sociobiology deals with the study of influence of Ecology and other biological processes on the social behavior of human being Aristotle argued that we are social animals; it could be argued that we are ecological animals, that we can fulfill our natural end only in a properly functioning ecosystem. This field of Sociobiology blended various field like biology, zoology of science with sociology.

Sociology is also treated as a critical science. For a variety of reasons, sociologists have tended to be critical of existing social organizations and institutions. Many sociologists have been visible advocates for dramatic social change. Their research has often been interpreted as an attack against the **status quo**. In the past, many sociologists have favored governmental intervention in order to insure more equity in society.

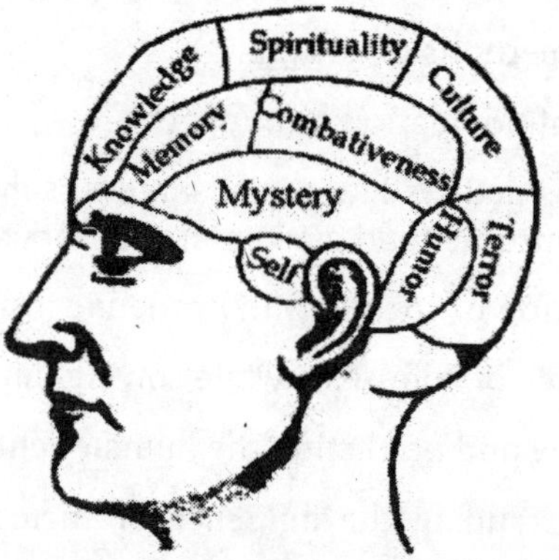

Sociology and Economics also share common field of interest as Economics deals with production of goods and their delivery. Whereas sociology deals with how efficiently the goods are delivered. Similarly Sociology and Political Science shares a common field of interest i.e. both are concerned with the welfare and upliftment of a human in a society. With popularity of statistical and mathematical formulation methods. It is being now become easier for Sociologists to calculate various output. Statistics contains the theoretical as well as practical aspects. It is used in analysing, evaluating and interpreting various facts and figures in addition to imparting knowledge, so that society and mankind get benefitted.

Social sciences are greatly influenced by diversity of factors which are the main cause of variations. Any social phenomenon and its impact on other phenomena's, their interdependence, their relationship with each other and their opposition to each other are only assessed and analysed through application of statistical tools. The process of census and demography are the most important application of statistics in sociology. Mortality rates, fertility rates, marriages and structure of population etc. are the result of application of statistical methods.

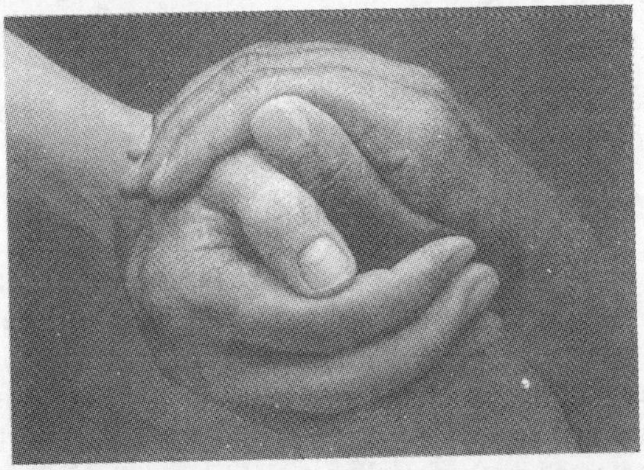

Social Science teaches us to how to act in a Society

Similarly with the evolution of computers, the mathematical or statistical problems are now become easier to be solved at the expense of little time. Computers also play a vital role in various sociological research works.

1.8 Sub field-disciplines of Sociology

The American Sociological Association has a list of Sections that illustrates the major interest areas. One critic argues that there are really 30 or 40 Sociologies "going off in every direction". Towards history, economics, politics, law, rural life, industry and religion. "There is no social activity that does not have its official sociologist". There are many Sub-fields in sociology. Here are some important ones:-

- Clinical
- Comparative
- Educational
- Environmental
- Industrial
- Legal
- Medical and health/wellness
- Political
- Religions
- Research design and methodology
- Rural
- Sociobiology (relations between genetics and social behavior)
- Sociolinguistics
- Sociology of art
- Sociology of Science
- Social Psychology (oldest sub field with focus on group dynamics, public opinion and socialization)
- Urban

In the late 1990s, Sociological theory, research methods and social organization appeared to be the most popular.

1.9 Industrial Sociology-Definition

Industrial sociology is a discipline which deals with study and examination of interaction of human in technological change, globalization, Labour markets, work organization, and managerial practices. Industrial sociology is also termed as sociology of work. Industrial sociology in general is an application and extension of various social facts and principles problems faced by a human being while working in any Industry. Industry includes all types of operational bodies like Academics, government, business and religious.

According to Miller and Form, "Industrial sociology is a substantive area of general sociology which might more accurately be termed the sociology of work organizations or the sociology of economy".

1.10 History/Development of Industrial sociology

(Suggestion: Teachers are requested to discuss the role of Industrial Sociology in classroom & elaborate this discussion with only some class room examples)

Idea of Industrial Sociology lies in the views of famous sociologist Karl Marx and Weber. The applied fields dealing with social behavior blossomed outside of sociology in 1950s/60s, at that time-Industrial sociology became a speciality in schools of business and management as well as some psychology department criminology and Family studies also experienced Substantial growth and became "independent" in some universities and colleges. During the same period sociologists became increasingly interested in data analysis and statistics. Demography, Social organization and social stratification became popular because more data available was available for research in these fields. Industrial sociology became much more focused on being scientific and there was less interest in social problems.

1.11 Scope of Industrial Sociology

With the development of industries, the requirement of sociologist and psychological group up a faster. They help in determine and defining the relations which exist in the field of management i.e. the relation between the Boss and workers, between the owner and Manager, between the managers of a Industries having. Several units, between the owner of different Industries, between the workers and many more.

These relationship study requirement increases the interest of researchers and the study helps in raising the output, solves various labour problem or we can say, it prevents the various conflicts at work place.

Industrial Sociology emphasizes more on factory workers, their working situations and especially on the technologies and various methods of production system employed. A part from this, it also deals with the factory outside mentality and relations of workers. Various comparative Surveys of different factories shows that the technologies are identical; this survey helps in defining the working levels of factories and their reactions on various problems like training, management, labours etc. had cleared a concept that Industries are generally affected by some complex politics, legal activities and some regulations in a Society and this also affects the national economic performance of a country.

1.12 Applications of Industrial Sociology

- It can be used to define various Industrial relations.
- It can be used for social development which entails Socio-political as well as economic changes.
- It can be used to define various Trade relations.
- It can be taught as post-graduate subject.
- Trained sociologists can be created and they may help in conducting various researches and training.
- It can be used as a source of trained personnel's for various social institutes.

- It helps in defining and determining the various possible relations inside or outside the country.

1.13 Labour Process theory

It is a branch of Industrial Sociology which deals with the act that how efficiently Labour power from workers is extracted by the Capitalists.

In 1974, Harry Braverman wrote a book named "Labour and Monopoly Capital: The Degradation of work in the twentieth century. This book provides a critical analysis of scientific management and analyzed Capitalist productive relations from a Marxist perspective.

Braverman, Harry (1920-1976)

A powerful theoretician and organiser became editor of The American Socialist until its demise in 1959. During the early 1960s he worked as an editor for Grove Press, where he was instrumental in publishing The Autobiography of Malcolm X. In 1967 he took over the editorship of Monthly Review Press, where he worked until his death. His most important work was the classic Labor and Monopoly Capitalism, which was published shortly before his death from cancer in August 1976.

Analysis of the labour process may be traced back to Karl Mark's interest in the means by which human labour is harnessed in the creation of products for human need. This process in seen to be socially organized and to vary historically between different modes of production under Capitalism, what appears as a relationship between things or objects in production, is in fact a social relationship between owners of the means of production and their work force. The key to understanding this relationship lies at the point of production in the management of the Labour process.

In Labour and Monopoly capital: The degradation of work in the twentieth century, Harry Braverman attempts to update this thesis, by an analysis of the Labour process in the era of monopoly Capital. This focus is on the so-called degradation of work associated with ever-tightening management control. It is postulated that a Subordination and de-Skilling of labour will emerge from the combined effects of modern management and new styles of mechanization and automation. The ideal management objective is the removal of all worker control or autonomy, to be achieved through the specialized division and sub division of tasks. Skilled craft-work is thus reduced to the status of unskilled Labour. Taylorism or scientific management which was developed at the beginning of the twentieth century is seen as the conscious and Systematic expression of this process of degradation. It is argued that one overall effect of the degradation of work will be to produce an affinity between intermediate-level workers and the mass of the working class.

Factor's which Braverman sees as bound up with the changes he postulates are state regulation of the economy, increasing emphasis on planning, the expansion of clerical work and office computerization and the emergence of dual Labour-markets. Precisely how these developments are linked to his central argument is not made clear-though this is also true of some related criticisms.

Braveman's conceptualization of the process of change has been queried in that it represents a unilinear trend, rather than a complex of factors, which may not necessarily cohere. There could, for example, be different patterns in different industries. Similarly, he is attacked for accepting one model of management as universal, when in fact there are many and varied strategies. However, Braveman's thesis has been contested by Andrew Fried man in his work "Industry and Labour" (1977). In it Friedman suggests that while the direct control of Labour is beneficial for the capitalist under certain circumstances, a degree of "responsible autonomy" can be granted to unionized or "core workers" in order to harness the skill under controlled conditions.

Also, Richard Edward showed in 1979 in "Contested Terrain", that although hierarchy in organizations has remained constant, additional forms of control (Such as technical control vie email monitoring, call monitoring, bureaucratic control via procedures for leave, sickness etc) has been added to gain the interests of the capitalist class versus the workers.

Other questions revolve around the nature and definition of skill, which is felt to allow more scruting than Braverman allows. The connection between the demise of craft skills and the process of rationalization is argued to be insufficiently explored, whilst the de-skilling hypothesis itself is apparently challenged by a dependence on newly emergent skills on the specific issue to under-emphasize the possibility of worker opposition to tightening management control, especially where there are strong trade unions; Whilst conversely, his equating of worker skill with weaker managerial control. has been challenged. This analysis is particularly limited since it addresses only the objective nature of class relations, not the subjective experience of the working class.

Summary

- Sociology is the scientific study of human social behavior, interaction between humans, social institutions and social organisations.
- Sociologists are divided into two groups– Micro Sociologist and Macro Sociologist.
- Micro sociologist looks at the particular interactions.
- Macro Sociologist looks at the pattern of interactions usually in larger groups.
- Some sub fields of Sociology are– Clinical, Educational, Environmental, Industrial Legal, Political, Religion, Comparative, Sociobiology and many more.

- Term Sociology was coined by Augusta Comte's cours de philosophic between 1830 to 1842.
- Sociologists play a role of counselor while solving minority or racial problems.
- Sociology maintains the closest relationship to other disciplines of science that also helps in studying about various aspects of society.
- Industrial Sociology is a discipline which deals with study and examination of interaction of human in technological change, globalization, Labour markets, work organization and managerial practices.
- Labour process theory is a branch of Industrial sociology which deals with the act that now efficiently Labour power from workers is extracted by the capitalists.

Objective type Questions

1. The word "Sociology" was coined by:-

   (a) Harry Braverman (b) Augusta Comte

   (c) Miller and form (d) Richard Edwards

2. The study of the interaction of people in the workplace is known as:-

   (a) Sociology (b) Sociobiology

   (c) Psychology (d) Industrial sociology

3. Which category of Sociologist looks at the pattern of interactions in larger groups?

   (a) Macro sociologist (b) Micro sociologist

   (c) Economical sociologist (d) Political sociologist

4. The study of the indentifying and comparing behavior of a particular group against the full range of human behavior:-

   (a) Anthropology (b) Sociology

   (c) Sociobiology (d) Psychology

5. The study of individual in any group or the relationship of group to one another i.e. its social attitudes and opinions:-

   (a) Anthropology (b) Sociology

   (c) Sociobiology (d) Psychology

6. The study of influence of ecology and other biological processes on the social behavior of human being:-

   (a) Anthropology (b) Sociology

   (c) Sociobiology (d) Psychology

7. The study and examination of interaction of human in technological change, globalization, labour markets, work organization and managerial practices:-
   (a) Anthropology (b) Sociobiology
   (c) Industrial sociology (d) Psychology

8. Labour Process theory seeks to explain:-
   (a) How Labours are managed?
   (b) How Labours are recruited?
   (c) How capitalists extract Labour power from workers?
   (d) Laws related to Labour problems

9. The Book "In Labour and Monopoly capital: The degradation of work in the twentieth century" was written by:–
   (a) Harry Braverman (b) Andrew fried man
   (c) Richard Edward (d) Karl Marx

10. The Book "Contested Terrain" was written by:–
   (a) Harry Braverman (b) Andrew fried man
   (c) Richard Edward (d) Karl Marx

Answer to objective Questions

1. b 2. a 3. a 4. a 5. d 6. c 7. c 8. c 9. a 10. c

Review Questions

1. Define the term sociology. How coined the term "sociology".
2. What is Anthropology?
3. Discuss the relationship of sociology with-
   (a) Psychology
   (b) Sociobiology
   (c) Statistics
4. Briefly discuss the development of Sociology.
5. Discuss about the scope of sociology.
6. Enlist the various sub-fields of sociology.
7. Enlist the applications of sociology.

8. Define the term Industrial Sociology. Cite examples which you have encountered while studying in classroom.
9. Discuss the scope and development of Industrial sociology.
10. Explain the Labour process theory.

Chapter 2

Rise and Development of Industry

2.1 Industrialization-An Introduction

Jean Fourastié
(1907-1990)

It is a process of changes including social and economical changes, through which the humans are switched from pre-industrial society to Industrial society. The socio-economic changes are closely related with technological innovations, particularly with the development of Energy source utilization technique and Mining of various earth metals like Iron, Aluminium etc. all these innovations are made for aim of achieving manufacturing standard.

According to Jean fourastie, An economy consist of three sectors- A "Primary sector" which deals with commodity production involving farming, livestock breeding and exploitation of mineral resources, A "Secondary sector" which deals with manufacturing and processing and A "Tertiary sector" which deals with different services. Industrialization process is historically based on the expansion of the secondary sector in an economy dominated by primary activities.

As per the views of Researchers, the factor's which facilates industrialisation and development are bounded from political-legal environments for industries, to full-fledge availability of cheap and skilled labours.

2.2 Early Industrialism- The Industrial Revolution

The Industrial revolution of the late 18th and early 19th centuries was revolutionary because it changed----revolutionized--the productive capacity of England, Europe and United States. But the revolution was something more than just new machines, smoke-belching factories; it increases productivity and increased the standard of living. It was a revolution which transformed English, European and American Society down to its very roots. The Industrial revolution serves as a key to the origin of modern western society. As Harold Perkin has observed "The Industrial revolution was no mere sequence of changes in industrial techniques and production, but a social revolution with social causes as well as profound social effects.

The industrial revolution can be said to have made the European working class. It made the European middle class as well. In the wake of the revolution, new social relationships appeared. As Ben Franklin once said "time is money". Man no longer treated men as men, but as a commodity which could be bought and sold on the open market. This "commodification" of man is what bothered Karl Marx-- his solution was to transcend the profit motive by social revolution.

There is no denying fact that the Industrial revolution began in England Sometime after the middle of the 18th century. England was the **"First Industrial Nation"**. As one economic historian commented in the 1960s. It was England which first executed "the take off into self-sustained growth". And by 1850, England has become an **Economic titan**. Its goal was to supply two-thirds of the globe with cotton spun, dyed and woven in the industrial centers of Northern England. England Proudly proclaimed itself to be the **"workshop of the world"**, a position that country held untill the end of the 19th century when Germany, Japan and United States overtook it.

More those the greatest gains of the Renaissance, the reformation, scientific revolution or Enlightenment, the Industrial revolution implied that man now had not only the opportunity and the knowledge but the Physical means to completely subdue nature. No other revolution in modern times can be said to have accomplished so much in so little time. The Industrial Revolution attempted to effect man's mastery over nature. This was an old vision, a vision with a history. In the 17th century the English statesman and **"Father of Modern Science"**, Francis Bacon (1561-1626), believed that natural philosophy (what we call Science) could be applied to the solution of practical problems, and so, the idea of modern technology was born. For Bacon, how could man enjoy perfect freedom if he had to constantly labour to Supply the necessities of existence His answer was clear--Machines. These labour saving devices would save labour which then could be utilized else where. **"Knowledge is power"** said Bacon and Scientific Knowledge reveals power over nature.

Sir. Francis Bacon (1561 - 1626). Later Baron Verulam and later still the 1st Viscount St Albans coined two major concepts: "Things alter for the worse spontaneously, if they be not altered for the better designedly" and "He that will not apply new remedies must expect and evils, for time in the greatest innovator. His comments on change and innovation are, of course, as valid in the 21st Century as they were 400 years ago.

Engines and machines, the glorious products of science began to revolutionize the idea of progress itself. While the Industrial Revolution brought its blessings, there was also much misery, Revolutions, Political or other wise, are always mixed blessings. If we can thank the Industrial Revolution for giving us fluoride, internal combustion engines, and laser guided radial arm saws, we can also damn it for the effect it had on social relationships. We live in the legacy of the Industrial revolution, the legacy of the **"Cash nexus"** as the mid-19th century Scottish Critic Thomas Carlyle (1795-1881) put it, where the only connection between men is the one of money, profit and gain. The elements needed or preferred for the Industrial revolution can be summarized as follows:

- Modern work altitudes
- Education
- A product
- Transportation for the product
- Large market
- Modern government
- Money

Thomas Carlyle (1795 – 1881)

Writer, historian, philosopher and literary critic. Born in Ecclefechan (Dumfries and Galloway). He wrote on a diversity of topics from the French Revolution to Oliver Cromwell. He was variously known as the *Prophet of Ecclefechan* and the *Sage of Chelsea.*

With innovations in manufacturing process started in England, it did spread to Europe and North America in the nineteenth century and to the rest of the world in the twentieth. This revolution not only did spread over the whole but it also imparted so many socio-economic political changes like:

- Wide and Equal distribution of wealth.
- International trade hike.
- Economic power of capitalist switched them to political power.
- Political power of capitalist influences greatly to the standard of peoples.
- Social changes like standard of working class and their relationships with their boss and other employees are also improved.
- For controlling the Labour power, managerial hierarchies were made.
- Various authorities according to the type of organisation are made.
- Environment friendly technologies were employed in order to cut the environment problems like pollution.
- Standard of living of middle class as well as Labour class was also improved.
- Free trade zone were made in order to enhance the economic state of any country.

2.3 Industrialization in Asia

The race of Industrialization firstly began in Japan in the late 19th country, then three more countries south Korea, china and India also joined this race in late 20th century. These four countries are known as the Asian tigers. These Asian tigers had an advantage of:

- Heavy Investments from other foreign countries.
- Minimized rate of custom duties.
- Cheap skilled workers.
- Competitive exchange rate.

- Well maintained and organised societies.
- Stable Government's existence.
- Free trade centers i.e. without any tariff trade centers.
- Abundance in availability of Labour force.

South Korea, one of the four Asian tiger, became economically powerful and wealthiest country of Asia due to its development from manufacturing of value added goods to steel, ship building and automobile industry in between the period of 1950s to 1980s and then to high-tech service industry in the 1990s & 2000s.

Japan-one of the four Asian tiger, opened the ports of Shimoda and Hakodate for International trade, this made Japanese government to reform its military and modernise it. In the 1870's the Japanese government promoted technological and industrial development as their base of Industrialisation. This lead Japan to became a powerful modern country.

India and China-two of the four Asian tigers were also following this pattern of development both of them are actively participating in expanding its own infrastructures. India is more concerned about its researches in fields like Nuclear technology, Information technology, Pharmaceuticals, Bioengineering and Technical education while chine was more concerned about increasing its military size and securing the energy and raw materials required with their investment in third world countries like Africa, they became Important player of World's economy.

2.4 Causes of Industrialization

The Industrial revolution began in England in the early 18th century for the following reasons:

- England had experienced all of the forerunners of industrialization in the previous century an agricultural revolution, cottage industry, and an expanded commercial revolution. These developments had built surplus capital and an infrastructure (shipping, banking, insurance, joint stock companies).
- England already had a handicraft textile industry using wool, used the availability of cotton from overseas markets as an alternative raw material.
- The Scientific revolution in England prepared the way for new inventions to be applied to industry like water frame, steam engine etc.
- A spreading shortage of wood (used for energy, for ship building and construction) stimulated a search for alternatives.
- England was rich in supplies of coal for energy and iron for construction.
- England had a long, irregular coast line with many rivers and natural harbors which provided easy transportation by water to many areas.
- England's population grew rapidly in the 18th century, providing a labor force for industry.

- Standard of Worker's were so miserable i.e. all the family members including men, women & children have to be involved for long hours.
- Skill of workers employed is also one of the reason that causes Industrialization.
- The accumulation of capital allowed investments in the concept and application of new technologies enabling the industrialisation process to continue to evolve.

2.5 The Industrial Revolution and Consequences

1. Towards a World Economy

Europe's entry into the Atlantic and Pacific created an international exchange of goods, created a new world-wide economy including the Americas, and paved the way for the establishment of colonies.

The ability of the West to dominate the seas allowed Western merchants to displace others from the world trade system. The Eastern particularly the so called Indian trade until the end of 15th Century was under the control of Islamic merchants.

Later at the end of the 16th Century the earliest of the colonizers, Spain and Portugal, declined in the face of later competition from England, France, and Holland..

The creation of a world economy largely dominated by the West was a major shift in history.

Latin America, Africa, the southern colonies of the American coast, and some other regions were drawn into a system that condemned them to an inferior, dependent status.

After 1700's, the West underwent a series of dramatic transformations in politics, intellectual development, and industrialization. The latter in the form of a so called "**revolution**" altered basic social patterns, enhanced Europe's position in the world, and marked the path to be taken by other civilizations.

2. "Water to the mill"

The answer to the question "where comes the water to the mill" ("değirmenin suyu nereden geliyor?" is a Turkish saying) that is the origin of the wealth of the West, lies mostly in the Atlantic slave trade.

Until 1630, the slave trade remained in the hands of the Portuguese. The Dutch and British began to export slaves to plantation colonies the America after 1637. Both Europeans and indigenous peoples were active participants in the commerce, because it was possible to realize major profits.

With the development of plantation agriculture in America, slaves became the primary component of the **coercive labor** system.

Slaves on ship moving towards Destination America

Between 1450 and 1850, about 12 million Africans were shipped to the plantations in America. Perhaps as many as four million more Africans were killed in enslaving wars prior to shipment.

The volume of slaves shipped increased from the sixteenth century to a zenith in the eighteenth century. By 1800, about three million slaves resided in the America. At its end in the nineteenth century, the slave trade still shipped more than one million slaves

High slave mortality in the plantation environment required constant replenishment of workers.

Slavery was part of the triangular trade, in which European manufactured goods were shipped to Africa for slaves sent to the plantation colonies from which sugar and cotton were exported to Europe.

In Africa, participation in the slave trade often reduced local economies to dependence on European manufactures. In this peculiar fashion, Africa was linked to the global trade system.

3. The Disruptions caused by "Industrialized" Life

Industrialization promoted movement from the country to the city. Family life was disrupted in the process.

Workers found themselves packed into slums and subjected to harsh labor conditions. Workers responded to new conditions, in some cases with outright resistance, but failed to slow the pace of technological change.

For the middle class, women retreated from the labor force to take up duties in the domestic household. Attitudes toward children involved greater concern for education and a sense of childhood.

A source of social instability was the dramatic population increase that set in after 1730, a result of improved nutrition and a lower rate of infant mortality. However unlike previous epochs, the population increase of the eighteenth century produced more positive responses.

Upper- class families attempted to secure their positions, and the social hierarchy became more rigid. Business families attempted to increase their margins of profit, by the addition of new technology.

At the lowest level, the poor were driven to seek new means of employment and the higher demand for work decreased the wages to a mere subsistence level.

Population growth stimulated a surge in production of textiles and other goods of daily usage. This process in history is called **proto-industrialization** accompanying changes in social patterns in response to new economic conditions

4. Towards the "Revolution"

During the improving stage of the market relations in European economies, the dominant mode was simple trade rather than production.

However, a manufacturing industry was also growing up around the logic of mercantilism. The arguments of the political economists of the period were similar to contemporary arguments. In order to have a favorable balance of trade it is more profitable to produce and sell goods with higher capital input and technology while importing cheap raw materials and agricultural goods.

The logic behind the new policy proposals was the growing conscience of domestic interests of a rising manufacturing bourgeoisie.

In a manufacturing economy, productive forces like capital and labor have to be employed in a way to constantly increase the productivity in order to increase the profits. This maximizing trend needs protection against external competition and needs a labor freely obtainable in the market.

Accordingly, this transition to an industrial economy needed more people freed from traditional bonds and obligations in the feudal domains in order to decrease the cost of labor..

We observe the preliminary signs of this trend in the 18th Century England. The English Parliament, unlike the monarchies of Europe, became more under the control of the merchant and capitalist classes.

Here we may remember the famous economist **Adam Smith**'s *The Wealth of Nations'* proposition that, "the only legitimate goal of national government and human activity is the steady increase in the overall wealth of the nation".

The key to wealth is surplus, and surplus is increased by producing more with less cost, then we arrive to the increasing drive for technological improvements and innovations. This nature of capitalism also explains the opposition of the bourgeoisie to feudal relations.

5. "Sheep are eating men": Enclosures

During the 17th and 18th Centuries, developments and inventions in science and technology applied increasingly to agriculture and manufactures.

Advances in farming and animal breeding methods increased productivity in the supply of raw materials, like wool, for the needs of the industry.

The term **"Enclosure"** is used to explain the process of appropriation of former common lands (lands which provided free access and use to the peasants) by the landlords to breed sheep for their wool.

Enclosures increased the number of larger and more productive and efficient capitalist farms but, reduced the peasants either to the status of farm laborers or more often displaced them from their lands.(Read the box: "Sheep are eating men")

During the whole 18th century growing number of peasants were obliged to migrate to towns and contributed the increase of cheap labor reserve for the manufacturers.

Urbanization accompanied the industrialization of societies, and the proportion of populations living in cities grew as societies were transformed. At the end of the 18th century, throughout the world the overwhelming majority of all populations lived outside of cities.

In Great Britain, the proportion of the population living in urban areas was 25 percent in 1831, became more than 50 percent in 1851, and had reached 77 percent by 1901.

Child Labour in Industry

6. Industry shouldered by children

The term "Industrial Revolution" signifies accelerated developments of the technology and their applications to the Industry of England. Supporting an unprecedented increase in industrial production, this revolution reached its peaks, at the end of the 18th century and mid 19th century.

Although by some authors the term "revolution" is considered to be exaggerated, it is true that between 1750's and 1850's the quality of goods produced increased enormously and machine power began to replace the human power for the first time in the history.

Workshops turned into factories, artisans to entrepreneurs and laborers to an alienated working class: the industrial proletariat. Output and distribution, passed the limits of local markets and the yields of a mass production began to be directed towards export markets.

Briefly, industrial "revolution" was not a sudden incident but a transformation through time resulting with technical,economic, social,political and ideological alterations.

If looked from another aspect, Industrial Revolution also would mean **misery** of the working people, including women and small children, laboring from early morning until night. In 1833 the English Government passed a *Factory Act* to improve conditions for children working in factories. Young children were working very long hours in workplaces where conditions were often terrible. The basic act was as follows:

- *Child workers under 9 years of age must possess a medical or age certificate*
- *Children between the ages of 9-13 to work no more than 9 hours a day*
- *Children between 13-18 to work no more than 12 hours a day*
- *Children are not to work at night*
- *Two hours schooling each day for children*

Four factory inspectors appointed to enforce the law throughout the whole of the country. However, the passing of this Act did not mean that overnight the mistreatment of children stopped.

The "Revolution" besides creating unsupportable working conditions was also the alteration of all urban and rural areas, the human and physical environment for the creation of industrial areas and workers' ghettos.

Child prisoner, 1870

This photo comes from an album showing prisoners in Oxford Gaol. The inclusion of photos in prison records at this early date is very rare. The child is Julia Ann Crumpling, aged seven. She was sentenced to seven days' hard labour on 28th June 1870 for stealing a baby carriage.

This type of punishment for a child convict was not all that unusual. Even though many people thought that sending a child to gaol did more harm than good, there were still over 1500 children in adult prisons in 1871. This was to change later in the century. After 1899, children had to be sent to separate places of correction, such as Reformatory Schools.

Industrial revolution also goes together with the resistance of worker in the form of machine breaking, which will be known as "**Luddite Movements**", rise of working class organizations, trade unions and finally "Socialism" as an ideology and movement.

New industrial techniques mostly originating in England during the early 19th Century spread rapidly in Northwestern Europe, France and the U.S.A. and during the second half of the 19th century to Germany, Italy and Austria.

Interestingly, these newly industrialized countries, particularly the U.S.A. as compared to England gained a better momentum towards the end of the century. This was due to the application of the most recent and efficient technologies without the transformation cost of outdated units. That is, countries industrializing later reached higher rates of development, while previously advanced countries due to the cost of replacing the old plants became relatively slower in the race of modernization.

The key development of the Industrial Revolution was the application of machine power to replace men and animals. Favorable supplies of natural resources and the spur of population growth helped to produce the first Industrial Revolution in Britain.

Industrialization built on the commercial advantages Europe enjoyed in the world trade network and the developments of the scientific revolution.

The initial inventions, such as James Watt's steam engine, that prompted the Industrial

Revolution occurred in Britain. Each invention spawned new technological developments in related fields.

James Watt's steam engine

Transportation and communication innovations allowed products, people, and information to be moved more rapidly. Improved agricultural production fed the masses of workers who moved to the cities.

Industrialization involved a shift in the organization of labor and the emergence of the factory system with its specialization of tasks and greater discipline. Industrialization also led to the creation of larger firms with greater access to capital and more advanced marketing techniques.

Britain's industrialization attracted imitators in the United States, Belgium, France, and Germany. The French Revolution promoted industrialization by sweeping away laws that restricted trade.

7. Transportation and communication

Developments in transportation were important aspects of the global processes of industrialization. Railroads made bulk transportation cheaper and easier, and railroad construction played an important role in industrialization.

The first railroads were built in Great Britain in the first quarter of the 19th century, and the first railroad was built in France in 1828. By 1850, rail networks had been established across western Europe and in the eastern United States.

The Berlin–Baghdad Railroad was an important scheme for the expansion of German economic influence in the Ottoman Empire.

Canals were also important in the transportation systems of emerging industrial societies; in fact, the construction of two major canals transformed the patterns of global maritime movements.

The Suez Canal, linking the Mediterranean Sea and the Red Sea (and ultimately the Indian Ocean) was completed in 1869 and reduced the need for west-east shipping to go around the African continent. It changed military and commercial shipping patterns and made Egypt a major global strategic location.

The Panama Canal was opened in 1914 and transformed strategic and commercial shipping in the Western Hemisphere. It joined the Atlantic and Pacific shipping routes in Central America, eliminating the need to go around South America (or through the Arctic regions).

8. Production technology

The technological innovations accompanied simultaneously the social and economic changes, and this may be considered as a reciprocal process.

The driving force of the Industrial revolution was the textile industry and particularly the cotton products.

Cotton industry was not considerable through much of the 17th century but exploded one century later.

Most cotton was produced in British colonies namely America and India. Because it was a labor- intensive agriculture, it fueled the traffic in African slaves to the American plantations.

The cotton produced in America imported to the homeland for the use of textile industry, and the finisheo product was marketed in Europe and in other countries.

At the beginning, having the technological superiority of production, England had almost the monopoly of cotton textiles. This position altered when America became independent and entered into the market as a competitor.

From our history we can also observe this competition which ended by the victory of America, so why since 1830's, the ordinary cotton textile is named “Amerikan bezi” in Turkey.

The first innovation in cotton manufacture was the **flying shuttle**, developed by John Kay in 1733, and was one of the key developments in weaving that helped fuel the Industrial Revolution.

The flying shuttle enabled the weaver to propel the shuttle through a wider strip of cloth with a single hand, and allowed the other hand to perform the combing to compact the cloth. This speeded the process and thus increased production.

John Kay's Flying shuttle

James Hargreaves, a carpenter, developed the spinning jenny, which greatly speeded up the process of weaving cotton threads into cloth. Patented in 1767, the spinning jenny was a series of simple machines rather than a single machine, and it spun sixteen threads of cotton simultaneously. (Picture below)

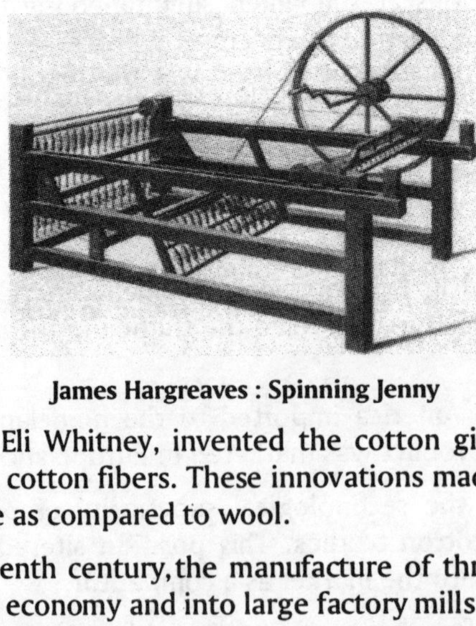

James Hargreaves : Spinning Jenny

In 1793, the American, Eli Whitney, invented the cotton gin, which mechanized the separating of seeds from cotton fibers. These innovations made cotton incredibly cheap and infinitely expandable as compared to wool.

By the end ofthe eighteenth century,the manufacture of thread and cloth was slowly moving out of the family economy and into large factory mills.

The invention that really drove the revolution in the eighteenth century was the steam engine.

Along with the growth in the cotton industry, the steel industry also had begun to grow.

This was largely due to the fact that England was rich in coal deposits, and this gave an advantage to England as compared to the countries of the continent.

In 1712, Thomas Newcomen built a simple steam engine that pumped water from the mines.(Below)

Newcomen Steam Engine

Later a Scotsman named James Watt added a separate cooling chamber to the machine in 1763; this cooling chamber condensed the steam so the cylinder itself didn't have to be cooled.

Patented in 1769, Watt's steam engine had the efficiency to be applied to all kinds of industries and began to change the face of English manufacture.

The nineteenth century saw the exporting of the Industrial Revolution to Europe and the explosion of factory-based, technology driven manufactures began to be the norm.

9. An overview of the 19th Century- Progress

The idea of "progress" of the Enlightenment that continuing growth and improvement being the destiny of human and natural life further consolidated during the 19th Century.

Charles Darwin's theory of evolution (Origin of Species, 1859) was taken as the confirmation that progress was the natural direction of life, and also valid in the analysis of social evolution. Herbert Spencer (1820 –1903) an English philosopher and liberal political theorist is considered as the father of Social Darwinism, a school of thought that applied the evolutionist theory of survival of the fittest (a phrase coined by Spencer) to human societies,

In economic theory liberals like David Ricardo (1772-1823) John Stuart Mill (1806-73) defended that free, unregulated competition would bring continuous economic expansion and establish the universal justice.

On the other hand Socialist theories arising from the injustices and inequalities caused by the industrialization, also based their ideas on human perfectibility and progress found its adherents among the working classes.

The new class of industrial workers responded by organizing trade unions, first legalized in England in1824. As a result of their efforts to reduce child labor and regulate conditions were passed in 1833.

From the cultural point of view, "**Realism**" in painting as well as literature appealed to a new mass audience with social or historical aspect of narrative.

With serious religious, moral, or social messages often drawn from ordinary life, the drawings of **Impressionists** rejected the formalism, sentimentality in favor of a spontaneous, un-detailed rendering of the world.

High population growth and "enclosure" movements transforming peasantry to "wage competing laborers" also reinforced this advantage.

Moreover, England's manufactured exports since the beginnings of the industrial revolution had been wool, cotton and hardware. If compared, for example to France's silk and other luxury goods, England's goods had a steady demand and a larger access to other markets.

The brief outline below explains the necessary conditions of a leap forward such as the "Industrial Revolution":

- Sufficient and accelerated capital accumulation.
- Availability and sufficient size of domestic and foreign market.
- Access and control of raw materials.
- Free and cheap labor.
- A political and institutional framework facilitating innovation.

England, as compared to the other countries was in a more advantageous state for the above mentioned preconditions.

During the 17th and 18th Centuries England had began to monopolize overseas trade in such a way that all its wars resulted with the acquisition of a new overseas territory.

Besides leading the North American, the South American and, most importantly, the Indian trades, England was also continuing to gain ground in the Mediterranean region.

An Impressionist forerunner: Monet's "Soleil Levant"

SOME IMPORTANT DATES ON TECHNOLOGY

1712 : The **Newcomen steam engine** is used to pump water from a coal mine.

1733 : John Kay invents the **flying shuttle** in textile machines.

1769 : Arkwright patents the **water frame** (weaving machine working with water power)

1776 : James Watt operates better version of Newcomen **Steam Engine**

1785 : Edmund Cartwright patents a **power loom**- weaving machine- with steam power

1800 : First **electric battery** invented by Volta.

1807 : Robert Fulton's **steamboat** Clermont, goes into service on the Hudson River.

1830 : George Stephenson's **locomotive**, The Rocket, operates between Liverpool and Manchester.

1836 : Samuel Morse invents the **telegraph.**

1840 : Samuel Cunard starts the regular **transatlantic passenger service** by steamships.

1842 : **Steam power** in a factory in the **Ottoman**

Empire, The Feshane in Defterdar, İstanbul. The use of the modern technology will also be operational in Beykoz deri in 1843, Basmahane, Bakırköy in1850.

1859 : Commercial **oil well** is drilled and starts to produce in Pennsylvania.

1855 : **Telegraph line** connecting **Istanbul** and **Edirne** inaugurated.

1860 : School for telegraph technicians opened in **Istanbul.**

1866 : The Siemens brothers ,Germany develop the **open-hearth process** of steel making.

1866 : First **transatlantic cable** for telegraph communication is laid.

1867 : First **railroad** transportation **in Turkey.** Izmir- Aydin line.

1867 : Alexander Graham Bell invented the **telephone.**

1879 : Thomas A.Edison invented **light bulb.**

1892 : Rudolf Diesel patented the **diesel engine.**

1899 : Guglielmo Marconi transmitted the first **wireless** message.

10. Health and Environment

Industrialisation has spawned its own health problems as it increases its adverse effect on natural resources as a result of which the ecological balance is disrupted. For example, the wastes of the industry produces various pollution like air, water pollutions, working of machinery produce unpleasant noise. The point of attraction is that our ecology is disrupted as human being is a part of ecology it also has its adverse effect due to Air, water and noise pollution. Various health problems like Tuberculosis, Deafness, Cholera Dysentery and many diseases are encountered. Long working hours of men, women and children also prevails many social and health issues. Inspite of many benefits, Industrialisation now became a major medical issue world wide.

2.6 Key innovations & Inventors of the Industrial Revolution

Technology, arguably the greatest aspect of the Industrial Revolution, can be simplified into a few different innovations and inventors, most inspired by one product. The first product to undergo the "revolution" from the cottage industry to the mechanized age was cotton. Britain, at the time, had a large wool trade. In 1760, the amount of wool exported was almost thirty times that of cotton. Demand for cotton grew with a change in the upper class fashion, and Britain started to allow more cotton production. Soon, not enough cotton could not be made to satisfy the demand. This demand was the inspiration for the following four inventions:

John Kay's "flying shuttle"

John Kay, a mechanic from Lancashire, patented the flying shuttle. Using cords attached to a picking peg, a single weaver, using one hand, could operate the shuttle on the loom. With this invention it took four spinners to keep up with one cotton loom, and ten people to prepare yarn for one weaver. So while spinners were often busy, weavers often waited for yarn. As such, the flying shuttle effectively doubled a weaver's production of cloth.

Weaving machine before and after implementing Kay's "flying shuttle" - Source

James Hargreaves' "spinning jenny"

In 1764, James Hargreaves invented the "spinning jenny," a device which allowed one person to spin many threads at once, further increasing the amount of finished cotton that a worker could produce. By turning a single wheel, one could now spin eight threads at once, a number that was later increased to eighty. The thread, unfortunately, was usually coarse and lacked strength. Despite this shortcoming, over 20 000 of the machines were in use in Britain by 1778.

Richard Arkwright's "water frame"

Also in 1764, Richard Arkwright created the "water frame" to produce yarn faster. The "Spinning-Frame," its earlier name, was too large to be operated by hand. After experimenting with other sources of power, he decided to employ the power of a water wheel, and his machine became known as the water frame. Rollers produced yarn of the correct thickness, while a set of spindles twisted fibers together. The machine was able to produce a thread far stronger than any other available at the time.

Samuel Crompton's "Crompton's mule"

In 1779, Samuel Crompton combined both the **"spinning jenny"** and the **"water frame"** to create a machine known as **"Crompton's mule,"** which produced large amounts of fine, strong yarn. With the arrival of these inventions, yarn had effectively become industrialized. By 1812, the cost of making cotton yarn had dropped by nine-tenths and the number of workers needed to turn wool into yarn had been reduced by four-fifths. The addition of these inventions to the work force moved the stress from the production to the supply of raw cotton. Within just a 35 year period, more than 100,000 power looms with 9,330,000 spindles were put into service in England and Scotland. Britain took advantage of the Americas' available new cotton, using it to help absorb the demand. By 1830, the import of raw cotton had increased to eight times its past rate and half of Britain's exports were refined cotton. At this point, the demand was high enough to provide inspiration for what is probably the most well known invention of the Revolution: the steam engine.

Samuel Crompton 1753-1827
Inventor of the Spinning Mule

Samuel Crompton revolutionized the cotton industry, but who received little personal gain for his efforts. Crompton was born at a Bolton farm called Firwood Fold, a keen and inventive mind as an accomplished mathematician had encountered Hargreaves' Spinning Jenny and Arkwright's Water Frame and felt that by combining elements of both (the rollers of the Water Frame and the twisting action of the Jenny), he could produce a much more effective machine. His first, hand-operated prototype was nicknamed the Spinning Mule, as it was a hybrid of the two earlier machines.
It made a huge difference to the spinning branch of the cotton industry. Not only did the Mule produce a strong and very fine yarn, but larger versions, powered by steam engines, could spin thousands of spindles at once. Indeed, many cotton magnets built factories especially to house these very long mules.

James Watt's "steam engine"

In 1769, James Watt patented the steam engine and in effect created a new source of power. Early-model steam engines were introduced to drain water and raise coal from the mines, but the crucial development was the use of steam for power. The first steam engine was actually produced by Thomas Newcomen, but Watt made it a reciprocating engine, creating the true steam engine.

James Watt
Mathematician and engineer whose improvements to the steam engine contributed to massive advances in the Industrial Revolution
(1736–1819)

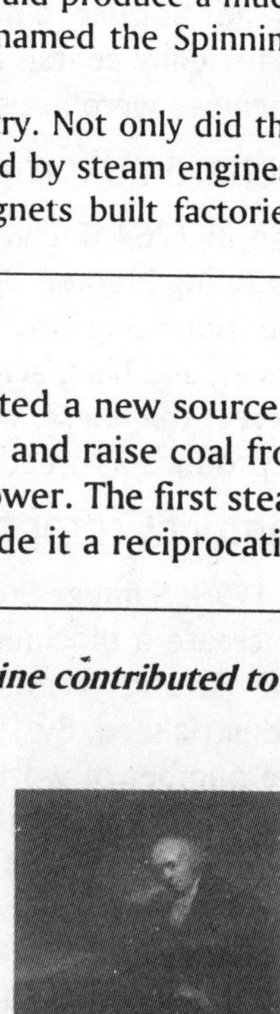

Watt solved the problem of how to convert the up-and-down piston movement to rotary movement (so that engines could power looms, bellows, and other mechanical devices in addition to extracting water from copper and tin mines as they were initially used).
He coined the term "horsepower" and nowadays his name is found written on almost every light bulb as the unit of electrical power.

Robert Fulton's "steamboat"

In 1807, Robert Fulton used steam power to create the first steamboat, an invention that would change the way and the speed in which materials could be moved between the colonies of Britain. In the beginning, the ship was more expensive to build and operate than sailing vessels, but the steamship had some advantages. It could take off under its own power and it was more steadfast in storms.

Stephenson's "steam powered train"

Stephenson's *Blucher*- Source

Finally, in 1814, Stephenson used the steam engine to create a steam powered train, which would eventually allow increased communication and trade between places before deemed too far. Britain encouraged the building of railroads in other European countries, often with British capital, equipment, and technicians. Railroads became a standard item of British export. From a suitable product comes a mass of inventions that will lead other areas of trade and production towards industrialization. These first innovations have greatly effected the basic elements of the era: agriculture, power, transportation, textiles, and communication.

The Textile Industry

The advancement of the textile industry was a key development in Britain's industrialization. It was consequently this industry that first employed the factory system. The raw materials used were essentially the same ones used under the domestic system, mainly featuring wool and cotton, but machines were now used to take the raw product and create fabric. With the use of machines and an "assembly-line" approach, it was possible to make enormous amounts of fabric in less time and for less money. Yet while advancements in this industry brought huge profits, and were therefore very good for the economy, there were many problems with how factories were run. Young children were employed, and were given exorbitantly small salaries. They were also forced to work extremely long hours in dangerous conditions, and were beaten in order to keep them working. It was not until the late 1820s that critics began to attack the way in which factories were run. Finally, in 1832, Michael Thomas Sadler headed a parliamentary committee on child labour known as the "Sadler Commission". But even when legislation designed to protect the worker was put in place, it was rarely implemented. Powerless workers eventually formed unions as a way of fighting the profit-hungry factory owners.

Inventions in the Textile Industry

1733 - Flying shuttle invented by John Kay - an improvement to looms that enabled weavers to weave faster.

1742 - Cotton mills were first opened in England.

1764 - Spinning jenny invented by James Hargreaves - the first machine to improve upon the spinning wheel.

1764 - Water frame invented by Richard Arkwright - the first powered textile machine.

1769 - Arkwright patented the water frame.

1770 - Hargreaves patented the Spinning Jenny.

1773 - The first all-cotton textiles were produced in factories.

1779 - Crompton invented the spinning mule that allowed for greater control over the weaving process.

1785 - Cartwright patented the power loom. It was improved upon by William Horrocks, known for his invention of the variable speed batton in 1813.

1787 - Cotton goods production had increased 10 fold since 1770.

1789 - Samuel Slater brought textile machinery design to the US.

1790 - Arkwright built the first steam powered textile factory in Nottingham, England.

1792 - Eli Whitney invented the cotton gin - a machine that automated the separation of cottonseed from the short-staple cotton fiber.

1804 - Joseph Marie Jacquard invented the Jacquard Loom that weaved complex designs. Jacquard invented a way of automatically controlling the warp and weft threads on a silk loom by recording patterns of holes in a string of cards.

1813 - William Horrocks invented the variable speed batton (for an improved power loom).

1856 - William Perkin invented the first synthetic dye (Bellis).

2.7 Types of Productive systems

The different types of Productive system's are as follows:-

(i) The Manorial or Feudal system

(ii) The Guild system

(iii) The Domestic or Putting out system

(iv) The Factory system

2.7.1 The Feudal System

The feudal system was introduced to England following the invasion and conquest of the country by William I (the conqueror). The system had been used in France by the Normans from the time they first settled there in about 900 A.D. It was a simple but effective system, where all land was owned by the King. One quarter was kept by the king as his personnel property, some was given to the church and the rest was leased out under strict controls. A simple plan showing how the feudal system works in given in Fig. 1.

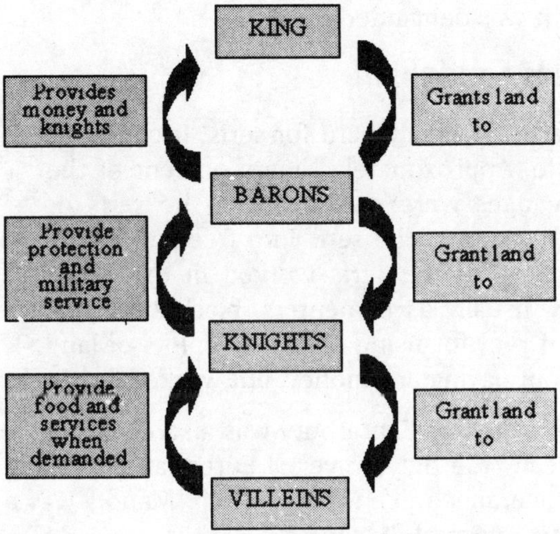

The King

The king was in complete control under the feudal system. He owned all the land in country and decided who he would lease land to. The men who leased land from the king were known as "Barons", they were wealthy, powerful and had complete controls of the land they leased from the king.

The Barons

Barons leased land from the king which was known as a Manor. They established their own system of justice, minted their own money and set their own taxes. In return for the land they had been given by the king, the Barons had to serve on the royal council, payment and provide the king with Knights for military when he demanded it. They also had to provide lodging and food for the king and his court when they traveled around the country.

The Knights

Knights were given land by Baron in return for military service when demanded by the King. They also had to protect the Baron and his family as well as the Manor, from attack.

Villeins

Villeins, sometimes known as serfs, were given land by knights. They had to provide the knight with free labour, food and service whenever it was demanded.

2.7.1.1 Life under Feudalism

The life on a manor was extremely hard for serfs. It consisted of work and family life. Approximately Ninety percent of the people in the middle ages were considered to be serfs or Villeins. There was a division of the serfs into free and a type of indentured servants. The free serfs worked in their own independent business, usually as carpenters, black smiths, weavers or backers. They paid the lord a type of rent for using their small plots of land. The other, in free serfs lives on the land without paying any money, but worked for the lord, earning their stay.

For example, the Archbishop of Canterbury was also chancellor of England in 1381. In fact the church was really the only universal European governing force. It was divided into spheres of influence, much like Manors. Each "Manor" was a diocese headed by a bishop. In addition to spiritual "Manors", many bishops were given real manors to govern. In this was, the church was firmly entrenched in the spiritual and practical lives of the medieval serfs. The church has a great influence over many of the common folk. The serfs believed that the harder they worked, the more of their money the gave to the church and the more they served the church, the better the after-life would be for them. The church also paid the lord to use the lord to use the land; and this sort of symbiosis between the church and the lord keep them both with an exceptional amount of money, while the serfs sometimes starved to death from over work and exploitation.

The term Manorilism was coined to determine and describe the political, economical and social system by which the serfs of medical period were bounded to their Land and lord. Manorilism in some times referred as the Feudal Society. This Manorilism system achieved its strength in

Eastern Europe after 15th century. While it declines by the 13th century after flourishing in 8th century in Western Europe.

The Manorilism i.e. feudal society was constructed for the reason of security i.e. The nobles wanted the security of maintaining control over their far-reaching kingdoms, so they were forced to delegate their power to local control. The serfs wanted security from marauders and barbarians from neighboring lands. They also wanted security from invading armies and thus the development of the feudal system and the manor structure was almost inevitable. But with the evolution of urbanization in Europe decline of feudal system started. This urbanization in Europe took the serfs from slavery towards learning new business and trade skills.

2.7.2 The Guild System

A guild is an association of craftsmen's who were involved in a particular trade. The earliest guilds in pre-industrial cities were formed as con fraternities of textile workers, masons, carpenters, carvers, glass workers, each of whom controlled secretes of traditionally imparted technology the 'arts' or 'mysteries' of their crafts. Guilds were founded by free independent Master craftsmen's.

2.7.2.1. History of Guild System

The origins of guilds remain obscure, as in their medieval form they appeared to combine characteristics the recall both the social solidarity of the Roman collegiums and a concern for skillful craftsmanship that is more easily identified with Germanic Societies in the early Christian centuries. Following the expansion of towns and trade after 1000 A.D., the social strata of the commune--a sworn association of equals developed into subgroups of people who practiced a common trade. This was particularly true in Europe's town and cities, although it must be seen in the context of an over whelmingly rural society. From the 13th century guilds became essential to most aspects of civic life and reflected the advance of the division of labour. They were responsible for, among other trades the production of leather and textiles (e.g. cloth merchants, dyers, stretchers, fullers, weavers, tailors, tanners and cobblers) and for building (e.g. masons, tillers, plasterers, carpenters and block smiths). In Nuremberg in the 13th century, for instance, there were craft organizations for gold smiths, cutlers, furriers, belt makers, cloth-weavers, armourers, sword smiths, scythe smiths, pewterers and mirror makers. Shortly after 1300, paper makers, wire smiths, bottle makers, brass-smiths, parchment makers, and bell-founders also formed guilds. Gunsmiths, glaziers, saddlers and carpenters soon followed and in 1400 the city council listed 141 separate crafts. Demarcation was strictly enforced: for example, veit stoss had great difficulty obtaining permission to cast a statue in bronze because the brass-smiths complained that this was not his trade. Approval for the project was eventually obtained through pressure from stoss's patron, the Emperor Maximilian.

Craft guilds ensured the maintenance of standards of production and of wage levels through their elected officers: a warden, a secretary, a treasurer and perhaps a committee. The guild was also responsible for the training that took the apprentice to the status of journey man and then master; work produced by the trainee in the workshop was submitted to a board of examiners, which consisted of qualified guild masters. Craft traditions were essential to the training of the artist; to become a master involved the production of a Master piece. The place of training and production for painters and for sculptors not working on site was the workshop.

Trade guilds were also religions organizations and many guilds met only as devotional confraternities. In England such devotional confraternities were often parish based. As Christian confraternities were under the protection of a patron saint, guilds met regularly for mass, for the celebration of a Saint's day and for an annual feast. They also played an important role in ensuring the burial of members and occasionally provided members' serving dependants with material help. Guild statutes make it plain that practical work and religions devotion were never clearly distinguishable. Attendance on the dead and provision for the washing of bodies were among the most significant responsibilities of guild officers, and occasionally there are macabre references in the documents to the need to empty a guild tomb.

2.7.2.2 Decline of the guilds

Despite of various advantages in agricultural and artisans field, at the end of 1700's and beginning of the 1800's, it became a target of much criticism. i.e. free trade had been opposed and several hindrance were paved down in the field of technological innovation, technology transfer and business development.

The factors responsible for decline of guild system are:

- The rights and trade area of concerned guilds was not defined properly.
- Negative roles of mediators.
- Replacement of guild system by factory system.
- Various inventions and discoveries opened and enlarged the area of trading which facilitates the guilds to produce products at maximum production rate.

2.7.2.3 Existence of Guild in Modern era

Guild in Modern era exists in different forms like.

- Real estate brokerage is an example of Modern American guild.
- In North America, The Newspaper Guild exists which in a Labor union for journalists and other newspaper workers.
- Various Law practices i.e. Bar association of every state in united states is also an example of modern guild.
- The Medical associations like American Medical Association and American Dental Association are also part of modern guild.

- Architecture, engineering, geology and Land surveying comes under the category of Professional guild.
- Commercial film makers, a collection of commercial, short film and feature film makers is an example of Guild in Australia.
- Various computer games players is an example of player guild.
- In united states, Quilting guild in very common and is found in almost all areas.
- Hollywood consists of many guilds like. The screen Actors Guild, The writer Guild of America East and The writer Guild of America, West.
- Consultants who work at many different companies and spreads new practices and knowledge between companies and corporations also defines a guild system.
- In the city of London, Livery companies which play a ceremonial role exists as a survivals of ancient guild.

2.7.3 The Domestic or Putting out System

The domestic system is sometimes called as Putting out system or Workshop system. In this system, the work was contracted to the contactors who complete their work with the help of their home members according to their own facility. Before 1750, most industry in Britain was small-scale. Most of it was literally "Manufactured" by hand-shoes, nails, knives etc. were made by craftsmen in small workshops usually at their homes. Most craftsmen worked from home, which is called the domestic system sometimes called cottage industry. The domestic or putting out system also gave rise to the terms like "merchant capitalism" where a craftsman finished work by number of other craftsmen. "Convenience agriculture" system where trader became a farmer or vice versa. The textile industry was the best example of the domestic system and varied between different parts of the country.

Home Craftsmen

In East Anglia and the west country, a clothier would give the work to craftsmen in their own homes-the clothier would buy the wool, take it to one home to be washed and carded, to another home to be spun, then to another worker to be woven. This kind of business is called merchant capitalism. Nevertheless, most work took place in worker's own homes. In the west Riding of Yorkshire, the system was different. There master clothiers worked for themselves. They would buy the wool at the market, then take it home where the whole family, perhaps helped by a few apprentices, would put it through the different processes under the same roof-the children would card it, the eldest daughter (a spinster) would spin it, the men would weave it. Then they would take the finished (10th to market to sell). Many craftsmen also owned Land. When trade was busy, they spent less time working on their form, and when trade was slack, they could spend more time on their farm. This system was called convenience agriculture.

Working from home

The picture shows a typical Domestic system home. The single room is dominated by a spinning wheel which is being worked by a young Lady-the spinster. Food is being cooked in the same room. A Ladder on the left of the picture will take the workers to their bedrooms once work for the day in finished and a window allows for light and ventilation. The amount of yarn produced in such a situation is clearly minimal.

If a worker did not work in his own home, he might work in a small work shop. Everything was done on a small scale. Even the coal mines to full local cottages rather than send coal further a field-were small with shallow bell pits being the favoured type of mine as opposed to deep coal mining.

Living and working

One of the most common goods introduced in people's homes was woollen cloth. As the population increased, there were more people to buy it! The whole family worked together to produce loads and loads of top-quality material. The whole process worked like this:

1 A **clothier** (wool tradesman) buys the wool from the farmer.

2 The clothier takes the wool to the villagers, who turn it into cloth.

FACT: Urghh!!

▸ The family collected their urine in a tub all week (what a smell!). It was later sold for a penny a tub to the local **fuller**, who used it to bleach cloth!

3 The family could work whatever hours they wanted ... as long as they met their deadlines.

4 The clothier collects the cloth and pays the villagers for what they have produced.
He or she brings them more wool for next week's order!

5 The clothier takes the cloth to be dyed, then sold.

It wasn't just wool that was turned into cloth. The soft fibres of the cotton plant (grown in warmer places like India) were turned into thread, which was then woven into cloth. Soon, cotton cloth became more popular than woollen cloth.

Domestic System

2.7.3.1 Characteristics of domestic System

- The workers involved could work at their own speed while at home or near their own home.
- Children working in the system were better treated in this system than they were to be in the factory system.
- As the women of a family usually worked at home, some one was always there to look after the children.
- Conditions of work were better as windows could be open, people worked at their own speed and rested when they needed to.
- Meals could be taken when needed, as people worked for themselves they could take a pride in what they did.
- Tension in the work place was minimal as the family worked as a unit.
- The best home produced goods were of a very good quality-though this probably was not true at a general level.

2.7.3.2 Draw backs of Domestic System

- The production was very slow and the finished product was simply not enough to, in the case of textiles.
- A better and faster system of production was needed, the complete process of production was usually done in several cottages and time was lost as materials were taken from cottage to cottage as one stage progressed to the next.
- The power of water was being developed and small cottages could not possibly take advantage of this source of power.
- Children's involved were so young to work in this domestic system.
- With a growing population that needed feeding. Clothing etc. a new way was needed to meet the demands.
- The industrial revolution marked the end of the domestic system,

When most people shifted homes to the factories as the centers of production of goods.

2.7.4 The factory system

It is a manufacturing method which was adopted first in England at the beginning of Industrial revolution. The factory system was a new way of organizing Labour made necessary by the development of machines which were too large to house in a Worker's cottage i.e. in domestic system. Further more, the efficient use of the new machines required that many of them be installed together where they could all be driven by the same power source.

Therefore, worker's homes become separately located from their place of work. All brought together under one roof it was considerably easier for the factory owner to supervise and closely regulate their workers.

Richard Ark Wright was the person, who was credited behind the growth of factories as he established a factory at cromford in Derby shire, England after patenting his water frame in 1767.

"The factory was more than just a larger work unit. It was a system of production resting on or characteristic definition of the functions and responsibilities of the different participants in the productive process".

David S. Landes, the unbound Prometheus

The Factory as a Machine

Individual machines as coordinated parts of larger, continuously operating mechanism

- **The mill clock - autonomous factory**

Many mills have a clock turned by the mill; close to another clock regulated by a pendulum, and the motion of the mill is so regular, that these two clocks will never vary more than two or three minutes. Both are made with dials and hands exactly alike, but one has a title on the dial, *mill* time, and the other, *clock* time.

THE MILL CLOCK

mill time, as measured by waterwheel Victorian clock from Pyemore Mill, near Bridport, Dorset

(J.M. Richards, *The Functional Tradition in Early Industrial Buildings*, 109 The Rees Cyclopedia, quoted by Hills, 193)

2.7.4.1 Comparison of Factory & Domestic System

Below given table shows some difference between the Domestic system and the factory system.

The Domestic System	The factory System
• All work was done by hand • Workers would receive the raw materials, take the home & build whatever was required & then return the finished product • Usually work was done in the Labourer's own home.	• The factory System was developed due to the advances being made in the textile industry. • With inventions hand weavers were driven out of business by big new factories, which they were later forced to work in. • These factories were first run by water, then by steam, and their output greatly improved the nation's economy. • Each worker concentrated on only one task . this "Assembly line" approach was very efficient. However the task becomes extremely monotonous and repetitive. • Poor working conditions led to Industrial Revolution but there were no Laws to protect workers and even when a few were passed they were rarely followed

Factory life: are you tough enough?

▸ What was it like to work in a factory in the nineteenth century?

If you could visit one of the first factories in Britain, perhaps the first thing you would notice would be the disgusting and dangerous working conditions. Most factory owners only cared about making a profit, not wasting money on providing a safe place to work. The machines were not fitted with any safety covers or guards and workers were not provided with goggles to protect their eyes. Factories were so noisy that people often went deaf and the dust made everyone sick. These hot, sweaty places would smell as well ... the stench would come from overflowing toilet buckets at the end of each room!

You would also notice a lot of children. Poor children did not go to school so boys and girls, some as young as five, would go to work with their parents. Put yourself in the position of a 14-year-old factory boy or girl in the early 1800s. They had probably been working for about seven or eight years. Look through the following information about working lives and ask yourself 'would you be tough enough to cope'?

▾ **Source A** From a report on conditions in factories made to Parliament in 1864

'Little boys and girls are seen at work at the machines and their fingers are in constant danger of being cut off. "They hardly ever lose a whole hand," said one of the owners, "it only takes off a finger at the first or second joint. They're just careless."'

▾ **Source B** From a modern history textbook published in the 1980s – 'History Alive 3, 1789–1914', by Peter Moss

'No one, not even small children, was allowed to sit down, except during lunch. Even during their short breakfast and tea breaks, many factories kept their engines running so that the workers had to eat with one hand and operate their looms or jennies with the other. These long hours spent hunched over the machines added stomach complaints, varicose veins and ulcers to the long list of diseases from which the workers suffered. Perhaps even worse, small children, who often had to bend their bodies in unnatural positions to do their job properly, frequently grew up with twisted spines, crooked thighs and knock-knees.'

▾ **Source C** An interview with Mr Moss, a factory supervisor, in 1816

Q: 'Were many children injured in the machinery?'

A: 'Very frequently, very often their fingers were crushed, and one had his arm broken.'

Q: 'Were any of the children deformed?'

A: 'Yes, several: there were two or three that were very crooked.'

Source D This source is based on an interview with a **pauper apprentice**. These were orphan boys who worked in the factories for no wages, but were given a bed and food to eat. The factory practically owned them.

'The room was full of young boys and girls all dressed the same, in just a shirt and trousers. We didn't wear shoes or socks. Dinner was hot: boiled potatoes, bread and porridge. Sometimes we had bacon too. We didn't get any plates; instead, we just held out the bottom of our shirts and they poured in the food. We were always so hungry that we ate every last bit. Next, the hungry crew ran to tables of newcomers and ate anything they'd left.'

Inspector: Tell me boy, where do you live?
Child: 36 Duke Street, Leeds.
I: Do you work in the factories?
C: Yes Sir.
I: At what age did you begin to work in them?
C: I was nearly eight years, I think.
I: How many hours a day do you work?
C: From six in the morning until seven at night.
I: Are you beaten at work?
C: Yes Sir. If we look up from our work or speak to each other, we are beaten.
I: If you don't go as fast as the machines, are you beaten?
C: Yes Sir. There's screaming among the boys and girls all day. They make black and blue marks on our bodies.
I: Are you allowed to 'make water' any time of the day?
C: No, only when a boy comes to tell you it's your turn. Whether you want to go or not, that's the only time you're allowed to go.
I: Can you hold your water for that long?
C: No, we're forced to let it go.
I: Do you spoil and wet yourself then?
C: Yes.

Source E An inspector's report on children in factories. What do you think is meant by the words 'make water'?

Source F Danger! None of the machines have safety covers or guards.

Source G A description of the accidental death of a young factory girl

'She was caught by her apron strings, which wrapped around the machine. She was repeatedly whirled around and around until she was killed. Her right leg was found some distance away from the machine.'

▾ **Source H** This set of rules appeared in a factory in 1844. The **overseer** was the workers' supervisor. 6d was about half a day's pay for a woman. Children earned about half the amount of women workers. Men earned twice as much as women.

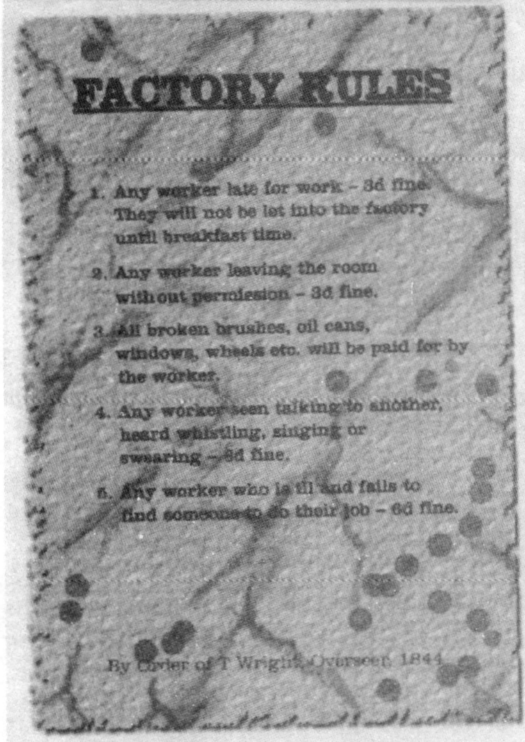

FACTORY RULES

1. Any worker late for work – 3d fine. They will not be let into the factory until breakfast time.
2. Any worker leaving the room without permission – 3d fine.
3. All broken brushes, oil cans, windows, wheels etc. will be paid for by the worker.
4. Any worker seen talking to another, heard whistling, singing or swearing – 3d fine.
5. Any worker who is ill and fails to find someone to do their job – 6d fine.

By Order of T Wright, Overseer, 1844

▾ **Source I** From a report to Parliament, 1843

'In Willenhall, West Midlands, the children are shamefully and cruelly beaten with a whip, strap, stick, hammer, handle, file or whatever tool is nearest to hand, or are struck with the clenched fist or kicked.'

▾ **Source J** From a leaflet about working conditions in 1837

'Girls loved their long hair, so they were sometimes punished by having it cut close to the head.'

▾ **Source K** An overseer about to punish a child. BEWARE: This picture appeared in a report on children's working conditions. The writer wanted Parliament to ban child labour. Some historians believe that the report, including this picture, was exaggerated to make things appear worse than they were.

▾ **Source M** A report written by a visitor to a mill in Bolton, 1844.

'The factory folk were better clothed, better fed and better educated than many other people. The factory was very clean. The working rooms were spacious, well ventilated and lofty, kept at an even temperature and clean. I observed great care in the boxing up of dangerous machinery and was told that accidents were very rare.'

Summary

- Industrial sociology is a process of changes including social and economical through which humans switched from pre - industrial society to industrial society.
- Industrial Revolution began in England sometimes after the middle of 18th century.
- England was the first Industrial nation and proudly proclaimed itself to be the workshop of the world.
- Race of industrialisation firstly began in Japan in the late 19th century then three more countries South Korea, China and India also joined this race in late 20th century.
- Japan, South Korea, China and India were known as Asian Tigers.
- The feudal System or Manorial system was introduced to England following the invasions and conquest of the country by William I(The conqueror)
- The Feudal system consists of four main elements: - The kings, The barons, The knights , The Villeins or serfs .
- A guild is an association of craftsmen's who were involved in a particular trade
- In the domestic system the work was contracted to the contractors who completed their work with the help of their home members according to their own facility.
- The factory system was developed in the late 18 the century chiefly due to advanced made being made in the Textile industry.

Objective Type Questions

1. Which Nation was the First Industrial Nation?

   (a) England (b) Germany

   (c) Russia (d) Japan

2. Which Nation proclaimed it self to be the "Workshop of the World"?

   (a) Germany (b) United States

   (c) Japan (d) England

3. Who was the Father of Modern Science?

   (a) Harold Perkin (b) Francis Bacon

   (c) Karl Marx (d) Thomas Carlyle

4. Which country among the given was known as one of the Asian Tigers?
   (a) South Korea (b) England
   (c) Germany (d) United states
5. Flying Shuttle was implemented in Weaving Machine by ----
   (a) Richard Arkwright's (b) John Kay's
   (c) Samuel Crompton's (d) James Hargreaves
6. Who invented the spinning jenny?
   (a) Richard Arkwright's (b) John Kay's
   (c) Samuel Crompton's (d) James Hargreaves
7. Who invented steam powered Train?
   (a) James Watt (b) John Kay's
   (c) Stephenson's (d) Robert Fulton's
8. Who invented Steam Boat?
   (a) James Watt (b) John Kay's
   (c) Stephenson's (d) Robert Fulton's
9. Who oppossed the complete control under the feudal system?
   (a) King (b) Barons
   (c) Knights (d) Serfs
10. Feudalism was charaterised by --
   (a) Vassals (b) Fiefs
   (c) Lords (d) All of the above
11. The Manorilism or feudal society was constructed for the reason of
   (a) Security (b) Trade
   (c) Slavery (d) Laws
12. Guilds are known to be precursors of -
   (a) Feudal System
   (b) Modern Corporation and Modern Trade System
   (c) Modern Courts
   (d) Modern Professions

13. Guild System was disbanded and replaced by free trade laws in -
    (a) Through 1800s
    (b) End of 1800 and beginning of 1900
    (c) In 1900
    (d) In beginning of 1800
14. In which system work was contracted to the contractors-
    (a) Domestic system (b) Guild system
    (c) Factory system (d) Feudal System
15. The new ways of organising labour was
    (a) Domestic system (b) Guild system
    (c) Factory system (d) Feudal system

Answer to objective Questions

1. a	2. d	3. b	4.a	5. b	6. d	7. c	8. d	9. a	10. d
11. a	12. b	13. a	14. a	15. b					

Review Questions

1. Define Industrialisation, Enumerate its various causes.
2. How phase of Industrialisation begins and its effect in Asia?
3. What are various consequences of Industrial Revolution?
4. Enlist the names of various inventors and inventions which lead to Industrial revolution.
5. What are the productive systems? Explain any two of them in brief.
6. Describe the Guild system and enlist the various forms of Modern Guild system.
7. Explain Manorial or Feudal system.
8. Explain Domestic or Putting system with its characteristics and drawbacks.
9. Describe factory system and state how it differs from domestic system.

Chapter 3

Industrialization in India

3.1 Introduction

The economic history of India begins with the Indus valley civilization. This civilization flourished between 2800 BCE to 1800 BCE. And most of the people at that time were involved in Agricultural activities and trade of tools, weapons made up of copper, bronze and tin was done. During 600 B.C., a trend of punch marked silver coins was started which involved in intensive trade and development of urban areas.

By 300 B.C. Era of Mughal empire started which united down many sub-continents of India as a result of which the trade route become more secure and reliable with the emergence of Mughal Empire in India, the agricultural productivity increased immensely due to free taxation rule for the farmers. Chandra gupta Maurya of Maurayan Empire, first time established the trend of single currency all across the India, this single currency rules facilitates justice and security to merchants, farmers and traders.

Between the period of 1st and 15th centuries A.D., India was estimated to have the largest economy and emerged as one of the fastest growing large economies of world.

3.2 Development of Industrialization in India

The Development of Industrialization in India is categorized under following stages

(i) Indus valley civilisation

(ii) Ancient and Medieval era

(iii) Maurya Empire

(iv) Mughal empire

(v) Era of Nawabs, Marathas and Nizams

(vi) British rule

3.2.1 Indus valley civilisation

This civilisation was flourished between 2800 BCE to 1800 BCE, this was the first known permanent and pre dominantly urban settlement which is a result of technological advanced and thriving economic system. The technological advance includes driven carts and boats. People of that civilisation were generally involved in agricultural, Animal husbandry and Weapon making activities. So, many evidences of this civilisation had been found in archeological research of cities like Mohenjo-Daro, Harappa, Lothal and Rakhigarhi, this research reveals their knowledge in Planning of streets, Layouts, Drainage system and Water supply. In other words we can say that their knowledge of urban planning.

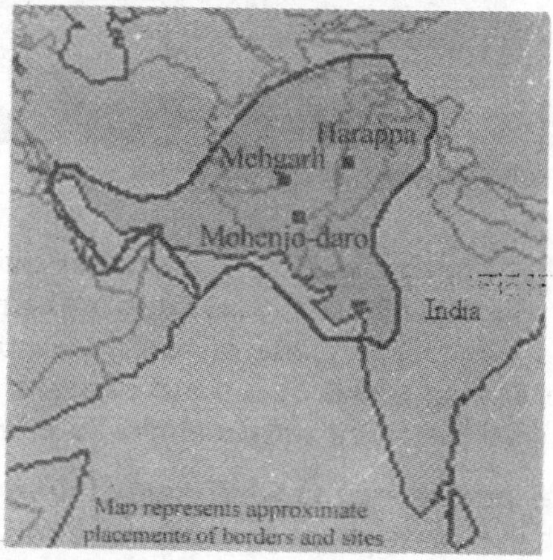

The Early Harappan 4,000 BC to 3,000 BC

Street in Mohenjo-Daro with Covered Drain

3.2.2 Ancient and Medieval Era

In this era, inspite of having significant urban populations, much of the populace resides in villages, where Agriculture was the predominant source of their economy. Agriculture provides food to the villagers as well as to the populace involved in other business like textile, food processing and crafts. Besides farmers some other class of people was there who were involved in activities like barberry, carpentry, weavers, smithy and Medical assistances.

Economic pattern of India at that time was much more similar to pattern of European guilds in which people are restricted from changing their occupation i.e. a barber cannot hold the work of smithy or Smither can't hold the work of Doctor. This guild system in India is a result of caste system, where personnel's belonging to a certain cast will have to perform work of that cast only like. Kshartiya's are allowed to involve in work of protection they can't do the work of Doctors and many similar cases were there. A part from this pattern, there was trends of joint family business i.e. younger's were trained in their business and after it they were employed in same parental business. This also imparts a drastic change in the economic system.

All to the economic historian Angus Madison mentioned in his book **"The world economy: A Millennial Perspective"** India was having Largest economy between 1st to 11th century. The credit of this achievement goes to Indian manufacturer who exported their specialized goods like shawls of Kashmir, cabicos of Bengal, pepper, cinnamon, opium, silk and Iron works to the countries of Europe, Middle East and South East Asia. This export gives them gold and silver in return.

3.2.3 Maurya Empire

This empire was flourished between 321 BC and 185 BC. At this time most of India was unified under one ruler for the first time. During this era the trade route was secure and risk associated with transportation reduces. Maurayan were much interested in infrastructure and Road system development. Chandragupta Maurya established a single currency across India which enhanced the trade. The Economic situation in Maurayan Empire and Roman Empire was some how similar they had extensive trade connections & both had organizations similar to corporations. The only difference between both empires was, Roman Empire had organizational entities, which were for public state commerce while Maurya Empire had private commercial entities for private commerce.

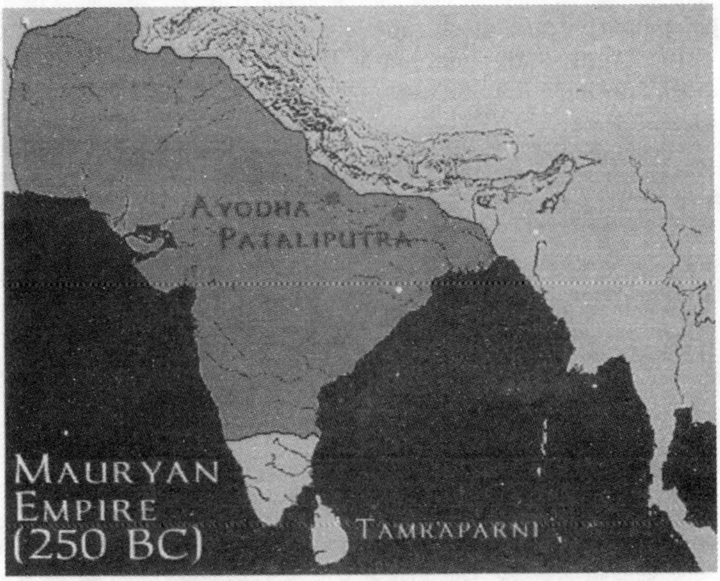

Map of the Mauryan Empire

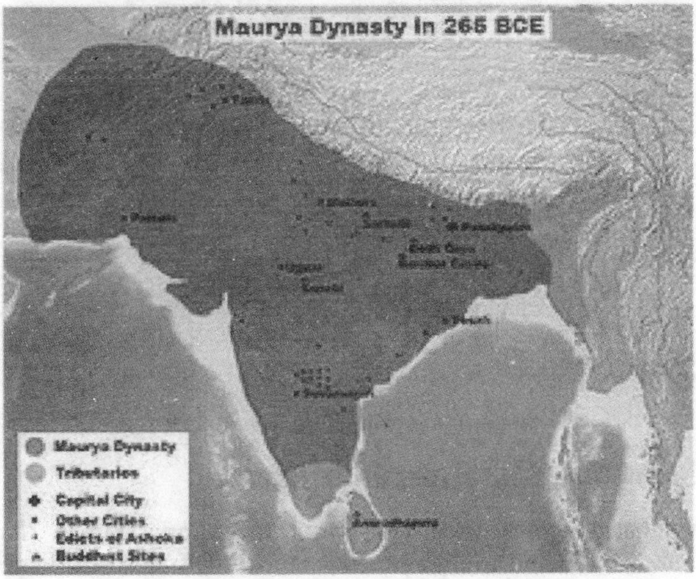

Maurya Empire under Ashoka the Great

3.2.4 Mughal empire

This empire flourished between 1526 and 1700. During this era India was the second largest economy in the world. By the 1700, the Mughal Empire expanded to almost 90 percent of South Asia with a uniform customs and tax administration system, at the time of Mughal Emperor Aurangzeb, annual revenue of India became twice that of European countries. Thus, India emerged as the Worlds' largest economy followed by china and Western Europe.

Mughal Empire in India, 16th century

3.2.5 Era of Nawabs, Marathas and Nizams

After fall of Mughal's empire, the Nawabs get their command in North India, the Marathas in Central India and the Nizams in South India. This era flourished between the periods of 1725 to 1775. In this era, the tax administration system was left largely intact, this made the India's economy to slip down and China became the World's largest economy followed India and Western Europe.

3.2.6 British rule

This era flourished between 1775 and 1847, the British rule in India was known as British East India was company rule they had created an institutional environment that stabilized the law and order to a very large extent. Britisher was more devoted towards the Infrastructure system development which includes railways, telegraphs and a modern legal system. These all strategy of development was mainly focused towards the exploitation of available resources in India. By the end of this rule, economy of India became one of the poorest economy is world as they were subject to frequent famines, had one of the world's lowest life expectancies suffered from pervasive malnutrition and were largely illiterate.

3.3 India on the path of development after Independence

India has made and secured an important position among other nations after independence especially in the field of expansion and diversification of production through Industry and agricultural work. More efforts were again necessary to day the foundation of development. This aim leads to reforms and revolunatory change in some of the following given sectors, which enhanced the standard level of India after freedom. These sectors are as follows:-

(I) AGRICULTURAL SECTOR

India is an agricultural country. Hence development in agriculture was indispensable for India's development. Even during the period of freedom struggle there were agrarian struggles in various parts of India the major agrarian struggles that later paved the way for the agricultural development of rural India are.

(a) Telangana Peasant Struggle

The peasants in the Telangana region of Hyderabad state suffered extreme feudal oppression at the hands of the Jagirdars and Deshmukhs.The peasants organised themselves against this. This was the basis of Telangana peasant struggle. By 1947-48 they could establish a firm base in the Nalgonda, Warangal and Khammam districts. Landlords fled to the towns, leaving their field free. The lands seized from landlords and the barren lands of the government were distributed to the peasants . This led to an encounter between the peasants and Indian army.

(b) Patiala Muzara Movement

At the time of India's independence the peasants of Patiala, the largest princely state in Punjab were in the midst of a struggle. After independence Patiala joined the Indian Union.

There were peasant struggles in many parts of India. It was because of the pressure of these struggles that the government tried to introduce land reforms. Moreover the Directive Principles of State Policy of Indian Constitution are against the concentration of wealth and means of production in the hands of a few. What does this indicate?

- Land in India was owned by a few

- Social justice and development of the country could not be achieved unless distributed among the peasants.

It was on the basis of this that land reforms were initiated after independence.

(c) Land Reform Acts

In 1949 the Central Government abolished the Jagirdari system. In the same year Zamindari abolition bills were introduced in a number of provinces such as U.P, Madhya Pradesh, Bihar, Madras, Assam and Bombay. This was challenged in the courts. The Congress government reacted by amending the constitution. The first amendment in 1951 and the 4th Amendment in 1955 were aimed at further strengthening the hands of the state legislatures for implementing Zamindari abolition.

In 1956 Zamindari abolition acts were passed in almost all provinces. The abolition of Zamindari meant that about twenty million erstwhile tenants now became land owners. The upper strata of the tenants who received tenancy directly from the landlords got more benefits. As the tenants had no title-deed, the land owners evicted them from the land.

(d) Tenancy Reforms

Tenancy system continued in various forms even though Zamindari system was abolished. Tenancy reforms were introduced to give ownership of land to the tenant who cultivated the land regularly. This benefited some peasants but all tenants did not get the right of ownership of land. The land owners and revenue officers subverted the rights of the tenants in many places.

(e) Land Ceilings

"In order to dispel the uncertainty regarding land reforms and ensure security to the farmer, ceilings should be fixed on the present and future holdings. Legislation to this effect should be completed in all states by the end of 1959". said the resolution of the Indian National Congress at Nagpur.

The states enacted different laws in order to determine the land ceiling. It was instructed that surplus land must be distributed among the landless people .But the majority of the landlords exploited the shortcomings of the legislation .and dodged its impact.

When the process of land distribution was obstructed, the peasants of Naxalbari in West Bengal and Srikakulam in Andhra Pradesh rose in revolt. This led them to an encounter with the state government Even though the struggles were suppressed it accelerated the speed of land distribution.

The successive efforts to determine land ceiling are.

- After 1971 the Central Land Reforms Committee asked the States to go through appropriate legislations to reduce the limit of land in individual possession.

- New laws were introduced in majority of the states. But all these were challenged by the landowners in the court.
- According to the 34th amendment of the constitution in 1974 these new legislations were included in the ninth schedule of the Constitution. Thus land ceiling laws became unquestionable in the courts. This helped to provide surplus land to many of the landless people

Bhoodan Movement

This movement was launched by Acharya Vinoba Bhave, The renowned Gandhian. Land reform was the objective of the movement. He gave shape to the Sarvodaya Sangh, an organization, to effect land reforms from outside the boundaries of law and to create a society free from violence.

(f) Agricultural Growth and the Green Revolution

After independence India depended on foreign countries for food.

India constructed several huge dams which laid the foundation of agricultural progress. Gather information on them and present in the class. Dependence on other countries for food will adversely affect the development of an agricultural country. So, Jawaharlal Nehru, Lal Bahadur Sastri and Indira Gandhi, our former Prime Ministers took initiative in bringing about basic changes in the agricultural field. Thus a new agricultural policy was framed its salient features are:

- Use of high quality seeds
- Use of fertilizers and machines
- Increase of irrigation facilities
- Facilities for soil testing
- Provision for agricultural loans

The age of Green Revolution comes after the implementation of land reform legislations which gave ownership of land to the peasant. In the absence of agriculture in the public sector the active participation of peasants was un avoidable for increase in production. Land reforms helped to equip them in this regard.

Thus we aimed at Green Revolution, a big leap in the agricultural field. Steps were taken to ensure fixed price for our agricultural products. This led to increase in food production. By 1980 India attained self sufficiency in food with a surplus of about 30 million to. India which imported food items before the Green Revolution began to export it.

(g) Origin of Co-operative Movement

We know about co- operative institutions. Such institutions came into existence as a result of the initiative taken by the national leaders after independence. Mahatma Gandhi and Jawaharlal Nehru believed that this will help the poor and the peasants to progress the impact of co-operative societies on the development of India are:

- Helped the poor peasants to get high quality seeds ,modern agricultural machines and low cost fertilizers .
- Helped the peasants to sell their products.
- Played a key role in making the Green Revolution a success

(h) Operation Flood

Agriculture and animal husbandry are mutually complementary and interdependent.

The Kheda District Co-operative Milk Producers Union Limited was formed in Anand, a small town in Gujarat. It helped the peasants of Gujarat to get a stable income. It was this Union which started the production of Amul, a brand of milk powder which became famous all over India. The success in Gujarat inspired other states to start similar co-operative societies. The National Diary Development Board was formed in 1965. By 1995 the peasants gained an income of about 3400 crores through co- operative diaries. The poor people were the main beneficiaries. One of the greatest successes in the history of Independent India, this is known as Operation Flood.

Operation Flood leading to White Revolution:

Operation Flood was a rural development programme started by India's National Dairy Development Board (NDDB) in 1970. The programmes objective was to create a nationwide milk grid.

It resulted in making India one of the largest producers of milk and milk products, and hence is also called the White Revolution of India.

Varghese Kurien (chairman of NDDB at that time), considered to be the architect of India's White Revolution (Operation Flood).

Operation Flood's objectives included:

Increase milk production ("a flood of milk")

Augment rural incomes

Fair prices for consumers

Operation Flood was implemented in three phases.

Phase I (1970-1980)

During its first phase, Operation Flood linked 18 of India's premier milksheds with consumers in India's major metropolitan cities: Delhi, Mumbai, Kolkata and Chennai, thus establishing mother dairies in four metros.

Phase II

Operation Flood Phase II (1981–1985) increased the milksheds from 18 to 136; 290 urban markets expanded the outlets for milk. By the end of 1985, a self-sustaining system of 43,000 village cooperatives with 4.25 million milk producers were covered. Domestic milk powder production increased from 22,000 tons in the pre-project year to 140,000 tons by 1989, all of the increase coming from dairies set up under Operation Flood. In this way EEC gifts and World Bank loan helped promote self-reliance. Direct marketing of milk by producers' cooperatives increased by several million litres a day.

Phase III (1985-1996)

Phase III (1985–1996) enabled dairy cooperatives to expand and strengthen the infrastructure required to procure and market increasing volumes of milk. Veterinary first-aid health care services, feed and artificial insemination services for cooperative members were extended, along with intensified member education.

Operation Flood's Phase III consolidated India's dairy cooperative movement, adding 30,000 new dairy cooperatives to the 42,000 existing societies organized during Phase II. Milksheds peaked to 173 in 1988-89 with the numbers of women members and Women's Dairy Cooperative Societies increasing significantly.

The **Women's Dairy Cooperative Leadership Programme** (WDCLP) was launched in 1995 as a pilot programme with the objective of strengthening the dairy cooperative movement by significantly increasing women's participation as active members and as leaders in the governance of cooperative societies, unions and federations.

Phase III gave increased emphasis to research and development in **animal health and animal nutrition. Innovations like vaccine for Theileriosis, bypassing protein feed and urea-molasses mineral blocks**, all contributed to the enhanced productivity of milch animals.

Operation Flood which started in 1970, concluded its Third Phase in 1996. Looking at what Operation Flood has achieved in milk, not simply at the application of science and technology, though both have played a role, not looking simply at the creation of farmer-owned structures, though such structures have been necessary to success, but at all of this, combined with the orchestration of all policies and programs that affect production. Further, they ensure to the extent possible, that these support mechanism strengthen efforts, rather than stand as obstacles.

Cost reduction and technology management

Modernization of process and plant technology

Interventions for productivity increase

Frontier technologies like DNA vaccines and genetically engineered bovine somatotropin, embryo transfer technology and in vitro fertilization of oocytes

(2) INDUSTRIAL SECTOR

Along with agricultural sector, the industrial sector also plays a major role in the development of a country. During the first few decades after independence there were no industrial magnates in India who were able to invest capital in large scale industries. At that time there were some limitations in receiving foreign capital .Foreigners were not interested in investing money in India as there were some crises that followed independence. So the public sector had to undertake the responsibility of industrial development.

Till 1985 we followed the development model devised by Nehru. It was decided that strategic industrial enterprises should be under the management of the government. It was the socialist idea that the controlling power of the economy should be in the public sector that led to this way of thinking. Hence large scale industrial enterprises were started under government ownership with economical and technological assistance from countries like Soviet Union, Britain, Germany and France. The basic necessities for the industrial development of any country are iron and steel industries and Petro chemicals industries. Prepare a list of such industries founded in India during the early years of independence

The achievements of the public sector during the early decades of Indian independence are:-

- The Public sector made great progress in industries like petroleum, steel, mining, electricity and petrochemicals.

- In agriculture based industries, manufacture of consumer goods and foreign trade the public sector played a remarkable role.
- The tendency of depending on import became minimized.

(3) EDUCATIONAL DEVELOPMENT

During the period of national movement detailed discussions were held about the future of Indian education system. National leaders including Gandhiji exhorted the people to boycott the educational institutions of the British. A number of national educational institutions were started during that period. The system of education that Gandhiji suggested for the whole of India is known as of Basic Wardha Scheme or Scheme Education. Find out the salient features of this education and present in the class.

Stages of Development

The Constitution amendment of 2001 declared that free and compulsory education of children of 6 to 14 years of age is a fundamental right.

This is not a sudden change. But an example of importance given to education by the governments of independent India.

The important education commissions appointed in free India for advising on education reforms are.

- The University Education Commission of 1948-49 under the chairmanship of Dr.S.Radhakrishnan.
- The Secondary Education Commission of 1952 under the chairmanship of Lakshmana Swamy Mudaliar.
- The Education Commission of 1964-66 under the chairmanship of Dr. D. S. Kothari.
- The Yashpal Committee under the chairmanship of Prof.Yashpal.

You might have heard of Navoday a schools opened as a part of the New Education Policy introduced in 1986. It was the Kothari Commission report which laid the foundation of this new policy. The Operation Blackboard scheme which provided for increased facilities in primary education and vocational education were started on the basis of new education policy. Media like radio and television began to be used for educational purposes. The State Council of Education Research and Training for promoting research and training in education at the state level and the District Institute of Education and Training in each district were started as a part of the New Education Policy. The commission led by Prof.Yashpal suggested measures for the improvement of the quality of education and simplification of work load.

The Minimum Level of Learning (MLL), District Primary Education Programme (DPEP) and the Sarva Siksha Abhiyan (SSA) were programmes devised to make basic changes in the field of primary education in India.

There were 27 universities and 578 colleges in India in 1951. Today there are about 250 universities and 11000 colleges in the country. Today we have medical colleges in the health education sector, engineering institutions including IITs and IIMs for management studies.

Many of these institutions are of international standard.

In order to control the various sectors of education and formulate policies, several national agencies are functioning in the state. They are as follows:

- University Grants Commission(UGC)
- National Council of Educational Research and Training (NCERT)
- All India Medical Council (AIMC)
- All India Council of Technical Education (AICTE)
- National Council of Teacher Education (NCTE)
- All India Bar Council. (AIBC)
- National Assessment and Accreditation Council (NAAC)

(4) DEVELOPMENT IN SCIENCE

In the beginning of the modern period the position of India in science and technology was not at all satisfactory. But with the beginning of the twentieth century considerable changes began to take place in this field. Jagadish Chandra Bose, Sreenivasa Ramanujan and many others led India to the path of modern science. After independence Indian scientists worked with devotion to attain scientific progress. The Scientific and Industrial Research Council founded in 1942 gained more prominence after independence. Some names of the scientists who gave leadership to the development of science in Independent India are:

(a) Dr.Vikram Sarabhai is regarded as the father of Indian Space Research Programme.

b) Homi.J.Bhabha gave leadership to research in atomic energy.

(c) A.P.J.Abdul Kalam, our President, is regarded as the father of Indian missile technology.

Space Research in India

The first rocket in India was launched from Thumba Rocket Launching Station, Thiruvananthapuram, on 21st November 1963. It was led by National Space Research Organization (INCOSPAR). Dr.Vikram Sarabhai was its head. The researches at Thumba paved the way for the launching of artificial satellites including INSAT.

Launching of P.S.L.V

Buddha Smiles

India established its nuclear potential with the atomic explosion at Pokhran in Rajasthan in 1974. 'Buddha smiles' was the code name given to it. With another atomic explosion at Pokhran in 1998 India declared herself an atomic nation.

Dr. Vikram Sarabhai : A scientist who dared to dream

Dr. Vikram Sarabhai's name will remain inseparable from India's space programme. It is well known that it was Dr. Sarabhai who put India on the international map in the field of space research. But he also made equally pioneering contributions in other fields such as textiles, pharmaceuticals, nuclear power, electronics and many others.

The resolution relating to the science policy of 1958 characterized science as the "Key to National Development". India's achievements in space research and missile technology are comparable to those of the developed nations. With the founding of Indian Space Research Organization (ISRO) on 15th August, 1969, India geared up for a big leap. The Space Commission came into existence in 1972.

India's achievements in the field of rocket launching have reached the level of the development of cryogenic engine. Rockets with cryogenic engines were necessary to amplify the thrust force so that the comparatively heavier communication satellites could be carried to the orbit. India is the sixth country in the world that possesses such satellite launching vehicles.

In missile technology India made achievements comparable to those of the developed countries. India had developed missiles for depending itself against the frequent foreign invasions what it had to face since independence. From missiles with short ranges to the ones with a long reach of 3000 Kms, India possesses missiles of all ranges.

(5) PLANNING IN INDIA

Planning in a democratic state is a social process in which, every citizen should have opportunity to participate. To set the patterns of future development is a task of such magnitude and significance that it should embody the impact of public opinion and the needs of the community.

We began to think of planning not after independence. The five year plans of the Soviet Union captured the attention of the leaders of India. Through the famous Karachi resolution of 1931 the Indian National Congress demanded that industries and service are as such as railway and water transport were to be in the public sector. In 1938 the National Planning Committee was formed under the leadership of Jawaharlal Nehru. This Committee prepared a plan outline for the all round development of India. In 1945 the famous Bombay plan, India's economic development scheme was drafted. This stressed the significance of the Public sector. It was in this background that the new Planning Commission came into existence in independent India.

The Planning Commission aimed at a mixed economy. India entered into the path of development through Five Year Plans the main achievements of the first three five year plans are:

- National annual production increased 4 times in half a century.
- Growth in the agricultural sector increased by 7.5% from the pre-independent era.
- There was 7.5% growth in the Industrial sector during the period along with development in the fields of education, health and science and technology.

India made these achievements at a time when foreign aid and private capital were meagre. This laid the foundation of India's development and the most important achievement of the period was the attainment of self sufficiency in food through Green Revolution.

Jawaharlal Nehru, the chairman of planning commission and Guljarilal Nanda, deputy chairman sign the draft of the third Five Year Plan.

Govt. of India accepted New Economic Policy in 1991 and its main features are:

- Commercial and Industrial regulations were liberalised.
- Import regulations were dropped.
- Aimed at the gradual privatisation of public sector.
- Industrial license policy was given up.

The restrictions on multinationals were lifted. Foreign investment was encouraged.

3.4 Synopsis on the impact of Industrialization upon various aspects of society

- Decline of sociability.
- Decline in family control.
- Decline in the influence of religion.
- Change in the institution of marriage.
- Change in institution of family.
- Change in the condition of women.
- Increase incidence of female infanticide.
- Commercializes solution of entertainment.
- Housing problem.
- Growth of Slums.

The Export became a reality, India became self sufficient with economic consequences of industrialization are:

- vast production of steel, cement, jute and sugar etc. with this vast production, India was self sufficient and start even export.
- Growth of trade.
- Division of labour and specialization.
- Economic crisis and unemployment.
- Industrial disputes and accidents.
- Problem of workers.
- Spread of socialism and individualism.
- Class conflict.
- Decline of rural industry.

3.5 Objectives of Industrial Policy

- To create additional employment opportunities.
- To create an environment for ensuring maximum value addition to the abundant, locally available mineral and forest based resources.
- To ensure balanced regional development by attracting industries in the economically backward areas of the state.

- To ensure participation of scheduled castes, scheduled tribes and other weaker sections in the development process.
- To make industrial investments in the state competitive vis-a-vis other states in the country.
- To promote private sector participation for creation of industrial infrastructure in the state.
- To create an environment for increasing industrial production, productivity and quality upgradation to face the challenge of competition emerging from economic liberalisation.

3.6 Strategies for imposing Industrial Policy

- To take measures for ensuring availability of necessary basic industrial infrastructure like rail-road, power, water, etc.
- To encourage private sector for providing quality roads, developed land, water, etc. in the least possible time, and to encourage captive power generation for providing low cost power.
- To establish new industrial areas, expand the existing industrial areas and to improve available services therein.
- To set up special industrial parks to provide common infrastructure and services by adopting cluster approach for the development of industries which have not been developed inspite of availability of abundant resources.
- To identify and promote such non-traditional industries for which good potential exists in the State due to availability of necessary resources.
- To provide special directed incentives to encourage establishment of industries by the weaker classes and in the economically backward areas of the State.
- To provide special incentives to small scale and cottage industries to create employment opportunities in the industrial sector in the least possible time throughout the State.
- To provide incentives to industrial units for technology upgradation and modernisation to enable them to face the challenges of global competition.
- To make arrangements for skill improvement, training and guidance of younger sections to enable them to seek maximum possible employment.
- To provide special package necessary for rehabilitation of sick and closed industrial units.
- To establish an effective system of "single point contact" and "time bound clearance" for providing requisite facilities, services and statutory clearances for investment.

3.7 Industrial Policy Resolution-1956

The Industrial policy resolution-1956 was shaped by the "Mahalanobis Model" of growth, which suggested that emphasis on heavy industries would lead the economy towards a long term higher growth plan. The resolution widened the scope of the public sector.

In this resolution industries are classified into three categories whose details are mentioned below:

(a) **Schedule A of the Resolution:** It comprises of 17 industries which are in direct domain of the Government and includes railways, airways, arms and ammunition, iron and steel and atomic energy.

(b) **Schedule B of the Resolution:** It comprises of 12 industries, which were envisaged to be progressively Sate owned but private sector was expected to supplement the efforts of the State.

(c) **Schedule C of the Resolution:** It contained all the remaining industries and was expected that private sector would initiate development of these industries but they would remains open for the State as well. It was envisaged that the State would facilitate and encourage development of these industries in the private sector, in accordance with the programmes formulated under the five year plans, by appropriate fiscal measures and ensuring adequate infrastructure.

The Resolution, in fact, reflected the prevalent value system of India in the early 1950s, which was centered on self sufficiency, in Industrial production. This policy was a landmark policy statement and was made to serve following objectives:

(i) To accelerate economic growth and boost the process of industrialization as a means to achieving a socialistic pattern of society.

(ii) Removal of regional disparities through development of regions with low industrial base.

(iii) Adequate infrastructure for industrial development of such regions was duly emphasized.

(iv) To provide all sorts of assistance to small and cottage industries for wider dispersal of the industrial base and more equitable distribution of income.

3.8 Industrial Policy Measures in the 1960s and 1970s

During this period of 1960s and 1970s, enormous number of policies were made in the path of development of Industries, like Monopolies Inquiry Commission (MIC) was set up in 1964 to review various aspects pertaining to concentration of economic power and operations of industrial licensing under the IDR Act 1951.

In 1967, the Industrial Licensing Policy Inquiry Committee (Dutt Committee) was setup, which recommended that larger industrial houses should be given licenses only for setting up industry in core and heavy investment sectors, thereby necessitating reorientation of Industrial licensing policy.

In 1669, the Monopolies and Restrictive Trade Practices (MRTP) Act was introduced to enable the Government to effectively control concentration of economic power. This Act defines large business houses as those with assets of Rs. 200 million and above. Large industries were designated as MRTP companies and were eligible to participate in industries that were not reserved for the Government or the small scale sector.

In 1970, New Industrial Policy was setup, which confined the role of large business houses and foreign companies to the core heavy and export oriented sectors.

This policy had categorized the industries into Four sectors which are as follows:

(a) **Core sector:** Consists of basic, critical and strategic industries.

(b) **Heavy investment sector:** Consists of projects involving investment of more them Rs. 50 million.

(c) **Middle Sector:** Consists of Projects involving investment between Rs 10 million to Rs 50 million.

(d) **De-licensed Sector:** Consists of Projects involving investment below Rs 10 million & this sector was exempted from licensing requirements.

3.9 The Industrial Policy Statement-1973

The Policy was set up in 1973 and named as Secretariat for Industrial Approvals (SIA). This policy gave preference to small and medium industrial houses over the large houses and foreign companies, in setting up of new capacity particularly in the production of mass consumption goods. All of the industrial houses are exempted from licensing requirements.

3.10 The Industrial Policy Statement -1977

This Statement emphasized decentralization of industrial sector with increased role for small scale, tiny and cottage industries. It also provided for close interaction between industrial and agricultural sectors. Highest priority was accorded to power generation and transmission. It expanded the list of items reserved for exclusive production in the small scale sector from 180 to more than 500. For the first time, within the small scale sector, a tiny unit was defined as a unit with investment in machinery and equipment up to Rs.0.1 million and situated in towns or villages with a population of less than 50,000 (as per 1971 census). Basic goods, capital goods, high technology industries important for development of small scale and agriculture sectors were clearly delineated for large scale sector. It was also stated that foreign companies that diluted their foreign equity up to 40 per cent under Foreign Exchange Regulation Act (FERA)

1973 were to be treated at par with the Indian companies. The Policy Statement of 1977 also issued a list of industries where no foreign collaboration of financial or technical nature was allowed as indigenous technology was already available. Fully owned foreign companies were allowed only in highly export oriented sectors or sophisticated technology areas. For all approved foreign investments, companies were completely free to repatriate capital and remit profits, dividends, royalties, etc. Further, in order to ensure balanced regional development, it was decided not to issue fresh licenses for setting up new industrial units within certain limits of large metropolitan cities (more than 1 million population) and urban areas (more than 0.5 million population).

3.11 Industrial Policy Statement - 1980

The industrial Policy Statement of 1980 placed accent on promotion of competition in the domestic market, technological upgradation and modernization of industries. Some of the socio-economic objectives spelt out in the Statement were i) optimum utilisation of installed capacity, ii) higher productivity, iii) higher employment levels, iv) removal of regional disparities, v) strengthening of agricultural base, vi) promotion of export oriented industries and vi) consumer protection against high prices and poor quality.

Policy measures were announced to revive the efficiency of public sector undertakings (PSUs) by developing the management cadres in functional fields *viz.*, operations, finance, marketing and information system. An automatic expansion of capacity up to five per cent per annum was allowed, particularly in the core sector and in industries with long-term export potential. Special incentives were granted to industrial units which were engaged in industrial processes and technologies aiming at optimum utilization of energy and the exploitation of alternative sources of energy. In order to boost the development of small scale industries, the investment limit was raised to Rs.2 million in small scale units and Rs.2.5 million in ancillary units. In the case of tiny units, investment limit was raised to Rs.0.2 million.

3.12 Industrial Policy Measures during the 1980s

Policy measures initiated in the first three decades since Independence facilitated the establishment of basic industries and building up of a broad- based infrastructure in the country. The Seventh Five Year Plan (1985-1900), recognized the need for consolidation of these strengths and initiating policy measures to prepare the Indian industry to respond effectively to emerging challenges. A number of measures were initiated towards technological and managerial modernization to improve productivity, quality and to reduce cost of production. The public sector was freed from a number of constraints and was provided with greater autonomy. There was some progress in the process of deregulation during the 1980s. In 1988, all industries, excepting 26 industries specified in the negative list, were exempted from licensing. The exemption was, however, subject to investment and locational limitations. The automotive industry, cement, cotton spinning, food processing and polyester filament yarn industries witnessed modernization and expanded scales of production during the 1980s.

With a view to promote industrialization of backward areas in the country, the Government of India announced in June, 1988 the Growth Centre Scheme under which 71 Growth Centers were proposed to be set up throughout the country. Growth centers were to be endowed with basic infrastructure facilities such as power, water, telecommunications and banking to enable them to attract industries.

3.13 Industrial Policy Statement- 1991

The Industrial Policy Statement of 1991 stated that "the Government will continue to pursue a sound policy framework encompassing encouragement of entrepreneurship, development of indigenous technology through investment in research and development, bringing in new technology, dismantling of the regulatory system, development of the capital markets and increased competitiveness for the benefit of common man". It further added that "the spread of industrialization to backward areas of the country will be actively promoted through appropriate incentives, institutions and infrastructure investments".

The objective of the Industrial Policy Statement - 1991 was to maintain sustained growth in productivity, enhance gainful employment and achieve optimal utilization of human resources, to attain international competitiveness, and to transform India into a major partner and player in the global arena. Quite clearly, the focus of the policy was to unshackle the Indian industry from bureaucratic controls. This called for a number of far-reaching reforms like:

- Abolition of industrial licensing for most industries, barring a handful of industries for reasons of security and strategic concerns, social and environmental issues. Compulsory licensing was required only in respect of 18 industries. These included, *inter alia,* coal and lignite, distillation and brewing of alcoholic drinks, cigars and cigarettes, drugs and pharmaceuticals, white goods, hazardous chemicals. The small scale sector continued to be reserved. Norms for setting up industries (except for industries subject to compulsory licensing) in cities with more than one million populations were further liberalised.

- Recognising the complementarily of domestic and foreign investment, foreign direct investment was accorded a significant role in policy announcements of 1991. Foreign direct investment (FDI) up to 51 per cent foreign equity in high priority industries requiring large investments and advanced technology was permitted. Foreign equity up to 51 per cent was also allowed in trading companies primarily engaged in export activities.

- Government provided automatic approval for technological agreements related to high priority industries and eased procedures for hiring of foreign technical expertise in order to inject technological dynamism in the Indian industry.

- Major initiatives towards restructuring of public sector units (PSUs) were initiated, in view of their low productivity, over staffing, lack of technological up gradation and low rate of return. In order to raise resources and ensure wider public participation PSUs, it was decided to offer its shareholding stake to mutual funds, financial institutions, general public and workers. Similarly, in order to revive and rehabilitate chronically sick PSUs, it was decided to refer them to the Board for Industrial and Financial Reconstruction (BIFR). The Policy also provided for greater managerial autonomy to the Boards of PSUs.
- Pre-entry scrutiny of investment decisions of MRTP companies was abolished. The thrust of policy was more on controlling unfair and restrictive trade practices. The provisions restricting mergers, amalgamations and takeovers were also repealed.

3.14 Industrial Policy Measures Since 1991

Since 1991, industrial policy measures and procedural simplifications have been reviewed on an ongoing basis. Presently, there are only six industries which require compulsory licensing. Similarly, there are only three industries reserved for the public sector. Some of important policy measures initiated since 1991 are set out below:

- Since 1991, promotion of foreign direct investment has been an integral part of India's economic policy. The Government ensured a liberal and transparent foreign investment regime where most activities are opened to foreign investment on automatic route without any limit on the extent of foreign ownership. FDI up to 100 per cent has also been allowed under automatic route for most manufacturing activities in Special Economic Zones (SEZs). More recently, in 2004, the FDI limits were raised in the private banking sector (up to 74 per cent), oil exploration (up to 100 per cent), petroleum product marketing (up to 100 per cent), petroleum product pipelines (up to 100 per cent), natural gas and LNG pipelines (up to 100 per cent) and printing of scientific and technical magazines, periodicals and journals (up to 100 per cent). In February 2005, the FDI ceiling in telecom sector in certain services was increased from 49 per cent to 74 per cent.
- Reservation for items that are manufacture exclusively in the small scale sector has been an important tenet of industrial policy. Realizing the increased import competition with the removal of quantitative restrictions since April 2001, the Government has adopted a policy of de-reservation and has pruned the list of items reserved for SSI sector gradually from 821 items as at end March 1999 to 506 items as on April 6, 2005. Further, the Union Budget 2005-06 has proposed to de-reserve 108 items which were identified by Ministry of Small Scale Industries. The investment limit in plant and machinery of small scale units has been raised by the Government from time to time. To enable some of the small scale units to achieve required economies of scale, a differential investment limit

has been adopted for them since October 2001. Presently, there are 41 reserved items which are allowed investment limit up to Rs.50 million instead of present limit of Rs.10 million applicable for other small scale units.

- Equity participation up to 24 per cent of the total shareholding in small scale units by other industrial undertakings has been allowed. The objective therein has been to enable the small sector to access the capital market and encourage modernization, technological upgradation, ancillarisation, sub-contracting, *etc.*
- In an effort to mitigate regional imbalances, the Government announced a new North-East Industrial Policy in December 1997 for promoting industrialization in the North-Eastern region. This policy is applicable for the States of Arunachal Pradesh, Assam, Manipur, Meghalaya, Mizoram, Nagaland and Tripura. The Policy has provided various concessions to industrial units in the North Eastern Region, *e.g.,* development of industrial infrastructure, subsidies under various schemes, excise and income-tax exemption for a period of 10 years, *etc.* North Eastern Development Finance Corporation Ltd. has been designated as the nodal disbursing agency under the Scheme.
- The focus of disinvestment process of PSUs has shifted from sale of minority stakes to strategic sales. Up to December 2004, PSUs have been divested to an extent of Rs.478 billion.
- Apart from general policy measures, some industry specific measures have also been initiated. For instance, Electricity Act 2003 has been enacted which envisaged to de-license power generation and permit captive power plants. It is also intended to facilitate private sector participation in transmission sector and provide open access to grid sector. Various policy measures have facilitated increased private sector participation in key infrastructure sectors such as, telecommunication, roads and ports. Foreign equity participation up to 100 per cent has been allowed in construction and maintenance of roads and bridges. MRTP provisions have been relaxed to encourage private sector financing by large firms in the highway sector.

Evidently, in the process of evolution of industrial policy in India, the Government's intervention has been extensive. Unlike many East Asian countries which used the State intervention to build strong private sector industries, India opted for the State control over key industries in the initial phase of development. In order to promote these industries the Government not only levied high tariffs and imposed import restrictions, but also subsidized the nationalized firms, directed investment funds to them, and controlled both land use and many prices.

In India, there has been a consensus for long on the role of government in providing infrastructure and maintaining stable macroeconomic policies. However, the path to be pursued toward industrial development has evolved over time. The form of government intervention in the development strategy needs to be chosen from the two alternatives: 'Outward-looking development policies' encourage not only free trade but also the free movement of capital, workers and enterprises. By contrast, 'inward-looking development policies' stress the need for one's own style of development. India initially adopted the latter strategy.

The advocates of import substitution in India believed that we should substitute imports with domestic production of both consumer goods and sophisticated manufactured items while ensuring imposition of high tariffs and quotas on imports. In the long run, these advocates cite the benefits of greater domestic industrial diversification and the ultimate ability to export previously protected manufactured goods, as economies of scale, low labour costs, and the positive externalities of learning by doing cause domestic prices to become more competitive than world prices. However, pursuit of such a policy forced the Indian industry to have low and inferior technology. It did not expose the industry to the rigours of competition and therefore it resulted in low efficiency. The inferior technology and inefficient production practices coupled with focus on traditional sectors choked further expansion of the India industry and thereby limited its ability to expand employment opportunities. Considering these inadequacies, the reforms currently underway aim at infusing the state of the art technology, increasing domestic and external competition and diversification of the industrial base so that it can expand and create additional employment opportunities.

In retrospect, the Industrial Policy Resolutions of 1948 and 1956 reflected the desire of the Indian State to achieve self sufficiency in industrial production. Huge investments by the State in heavy industries were designed to put the Indian industry on a higher long-term growth trajectory. With limited availability of foreign exchange, the effort of the Government was to encourage domestic production. This basic strategy guided industrialization until the mid-1980s. Till the onset of reform process in 1991, industrial licensing played a crucial role in channeling investments, controlling entry and expansion of capacity in the Indian industrial sector. As such industrialization occurred in a protected environment, which led to various distortions. Tariffs and quantitative controls largely kept foreign competition out of the domestic market, and most Indian manufacturers looked on exports only as a residual possibility. Little attention was paid to ensure product quality, undertaking R&D for technological development and achieving economies of scale. The industrial policy announced in 1991, however, substantially dispensed with industrial licensing and facilitated foreign investment and technology transfers, and threw open the areas hitherto reserved for the public sector. The policy focus in the recent years has been on deregulating the Indian industry, enabling industrial restructuring, allowing the industry freedom and flexibility in responding to market forces and providing a business environment that facilitates and fosters overall industrial growth. The future growth of the Indian industry as widely believed, is crucially dependent upon improving the overall productivity of the manufacturing sector, rationalisation of the duty structure, technological up gradation, the search for export markets through promotional efforts and trade agreements and creating an enabling legal environment.

3.15 Negative List of Industries Units Ineligible For Exemption / Concession

- Ice factory, manufacturing of ice cream, ice candy and Ice fruit
- Confectionery, biscuit and bakery products (excluding certified packaged and branded products obtained from mechanised process)
- Manufacturing of sweets, gazak and rewadi
- Manufacturing of namkin, purification of edible salt (excluding standardised packaged and branded products)
- Grinding of masala and chillies, manufaturing of papad (excluding standardised packaged and branded products)
- Flour mill (excluding Roller flour mill)
- Huller mill
- Book binding, manufacturing of envelopes, paper bags, playing cards and paper cone
- Saw mill, all types of wooden items, carpentry, wooden furniture (excluding wooden handicraft)
- Cloth / paper printing press (excluding handicraft printing and offset printing)
- Manufacturing of bricks, ridges (excluding fly ash bricks, fire bricks and brick manufacturing from mechanised process)
- Tyre rethreading (job work)
- Stone crusher , manufacturing of ballast (gitti)
- Coal briquette, coke and coal screening , coal fuel
- Powdering of mineral (excluding standardised branded products)
- Manufacturing of lime powder, lime chips, dolomite powder, mineral powder and Lime
- Lamination (excluding lamination of jute bag)
- Electrical job work
- Soda / mineral / distilled water (excluding standardised branded products)
- Preparation of pan masala, supari, tobacco, gutkha
- Manufacturing of crackers, Aatishbaji
- Repacking of goods
- Blending and packing of Tea (excluding standardised branded products)

- Photo laboratories
- Soap and Detergent (excluding standardised branded products)
- All types of Coolers
- Photocopying and Stenciling
- Rubber stamp making
- Bardana repairing
- Polythene bags (excluding H.D.P.E.)
- Leather tannery
- Public undertakings of Government of India or any State Government (excluding joint undertakings with private companies)
- Such other industries as notified by the State Government

3.16 List of Special Thrust Sector Industries

- Processing of medicinal, aromatic and dye plants
- Automobile, auto components, spares and cycle industries
- Manufacturing of plant, machinery & engineering spares
- Downstream products based on aluminium
- Food processing industries (industries approved for subsidy / assistance from Government of India)
- Milk chilling plant and branded dairy products
- Pharmaceutical industry
- White goods and Electronic consumer products
- Power generation from non-conventional sources
- Information technology, bio technology and advanced technology industry
- Such other industry as may be notified by the State Government

3.17 List of Industries Eligible for Exemption from Stamp Duty

- In the case of small scale industry, excluding the industries in Annexure–2, all industries will get the exemption.
- Medium – Large Industry – Mega Project : Following industries will get the exemption –

- Herbal and medicinal plant processing industries
- Automobile, auto components, spares and cycle industries
- Manufacturing of plant / machinery / engineering spares
- Downstream products based on Aluminium
- Food processing (industries approved by the Government of India for subsidy / assistance)
- Milk chilling plant and branded dairy products
- Pharmaceutical industry
- White goods and electronic consumer products
- Power generation by non-conventional sources
- Information technology, bio technology and advanced technology
- Other thrust sector industries
- Forest produce based processing industries
- Iron and steel, and industries based thereon
- Cement and cement based industries
- Coal and chemical industries
- Precious stones and ornaments
- Granite based industries
- Road and urban infrastructure including development of new Raipur city
- Water supply
- Energy generation, transmission and distribution
- Rice bran oil solvent extraction plant
- Board and paper mill based on paddy waste
- Cold storage
- Lemon grass oil, menthol oil
- Bamboo based paper industry
- Flower based ayurvedic medicine manufacturing
- Flower based scents and perfumes
- Such other industries as may be notified by the State Government

3.18 Industrial licensing policy

Industrial Licensing is governed by the Industries development & Regulation Act, 1951. The Industrial Policy Resolution of 1956 identified the following three categories of industries:

- Those that would be reserved for development in the public sector;
- Those that would be permitted for development through private enterprise with or without State participation; and
- Those in which investment initiatives would ordinarily emanate from private entrepreneurs.

Over the years, keeping in view the changing industrial scene in the country, the policy has undergone modifications. Industrial Licensing policy and procedures have also been liberalised from time to time. A full realisation of the industrial potential of the country calls for a continuation of this process of change.

In order to achieve the objectives of the strategy for the industrial sector for the 1990s and beyond, it is necessary to make a number of changes in the system of industrial approvals. Major policy initiatives and procedural reforms are called for in order to actively encourage and assist Indian entrepreneurs to exploit and meet the emerging domestic and global opportunities and challenges. The bedrock of any such package of measures must be to let the entrepreneurs make investment decisions on the basis of their own commercial judgment. The attainment of technological dynamism and international competitiveness requires that enterprises must be enabled to swiftly respond to fast changing external conditions that have become characteristic of today's industrial world. Government policy and procedures must be geared to assisting entrepreneurs in their efforts. This can be done only if the role played by the Government were to be changed from that of only exercising control to one of providing help and guidance by making essential procedures fully transparent and by eliminating delays.

The winds of change have been with us for some time. The industrial licensing system has been gradually moving away from the concept of capacity licensing. The system of reservations for public sector undertakings has been evolving towards an ethos of greater flexibility and private sector enterprise has been gradually allowed to enter into many of these areas on a case by case basis. Further inputs must be provided to these changes which alone can push this country towards the attainment of its entrepreneurial and industrial potential. This calls for bold and imaginative decisions designed to remove restraints on capacity creation, while, at the same time ensuring that overriding national interests are not jeoparadised.

3.19 Foreign Investment

While freeing Indian industry from official controls, opportunities for promoting foreign investments in India should also be fully exploited. In view of the significant development of India's industrial economy in the last 40 years, the general resilience, size and level of sophistication achieved, and the significant changes that have also taken place in the world industrial economy, the relationship between domestic and foreign industry needs to be much more dynamic than it has been in the past in terms of both technology and investment. Foreign investment would bring attendant advantages of technology transfer, marketing expertise, introduction of modern managerial techniques and new possibilities for promotion of exports. This is particularly necessary in the changing global scenario of industrial and economic cooperation marked by mobility of capital. The Government will therefore, welcome foreign investment, which is in the interest of the country's industrial development.

In order to invite foreign investment in high priority industries, requiring large investments and advanced technology, it has been decided to provide approval for direct foreign investment upto 51 per cent foreign equity in such industries. There shall be no bottlenecks of any kind in this process. This group of industries has generally been known as the "Appendix I industries" and are areas in which FERA companies have already been allowed to invest on a discretionary basis. This change will go a long way in making Indian policy on foreign investment transparent. Such a framework will make it attractive for companies abroad to invest in India.

Promotion of exports of Indian products calls for a systematic exploration of world markets possible only through intensive and highly professional marketing activities. To the extent that expertise of this nature is not well developed in India, Government will encourage foreign trading companies to assist us in our export activities. Attraction of substantial investment and access to high technology, often closely held, and to world markets, involves interaction with some of the world's largest international manufacturing and marketing firms. The Government will appoint a special board to negotiate with such a firm so that we can engage in purposive negotiation with such large firms, and provide the avenues for large investments in the development of industries and technology in the national interest.

3.19.1 Foreign Technology Agreements in India

The **Foreign Technology Agreements in India** plays an important part in the technological development of the Indian industries. The agreements are permitted through Government approval or by RBI governed automatic route.

Foreign Technology Agreements in India caters to the growth of the technology in the Indian industries. The primary purpose of the agreement is to induce the required amount the technological development and promotion of technologically advanced industries. The foreign technology is transferred from foreign sources such as research and development agencies, foreign parent companies, and other sources to the Indian counterparts. The transfer of foreign technology takes places by the means of foreign direct investments and foreign technology collaboration agreements.

Foreign Technology Agreements in India permits transfer of technology by the means of Government approval or through the automatic route delegated by RBI.

Automatic Approval-Foreign Technology Agreements in India

- The RBI grants automatic approval by the means of the regional offices to Indian industries for foreign technology collaboration.
- The payments pertaining to the technology transfer should not exceed US$ 2 million
- The royalty to be paid is restricted to 5 % in case of domestic sales, 8 % in case of exports and total payment should be 8 % on sales for a period of 10 years
- The royalty period should not exceed 7 years from the date of starting of the business or 10 years from the date mentioned in the agreement
- The royalty paid is 1 % in case of domestic sales and 2 % in case of exports as

granted under automatic route for use of the trademark foreign collaborator

- The royalty paid is 5 % in case of domestic sales and 8 % in case of exports as granted under automatic route for wholly owned subsidiaries

Government approval is required in the following cases -

- Proposals which requires licensing
- The manufacturing and products should be compliant with the small scale industries
- Proposals which involves previous trademark agreement, joint ventures, technology transfer, etc
- In case of an extension of the foreign technology collaboration agreements which had been automatically approved earlier
- Proposals with no parity with the norms of the automatic approval.

3.19.2 Foreign Technology Collaboration in India

The main aim of the Foreign Technology Collaboration in India is its technological development and enhances the productivity rate and its efficiency.

Foreign Technology Collaboration in India permits transfer of technology by the means of Government approval or through the automatic route delegated by RBI. The collaboration induces the required amount of technological development and promotion of technologically advanced industries. Foreign Technology Collaboration in India is undertaken with the objective of improving technology levels in the Indian industries. This helps to increase the efficiency and productivity of the industries in India.

Foreign Technology Collaboration-Methods of approval

The methods of approval Foreign Technology Collaboration in India helps in effective transfer of technical aspects. The transfer of foreign technology plays an important role in the Foreign Direct Investments. With the introduction of the 1991 Economic Policy, foreign investments started inflowing in large amounts. The Central government has allowed the transfer of technology in various industrial sectors. Significant technology transfers have taken place since the economic reforms.

The Central Government presented the Statement on Industrial Policy in 1991, which provided simplified methods for the better governance of the Foreign Technology Agreements. Sec 39C deals with Foreign Technology Agreements. There are certain standardized clauses for the approval of the Foreign Technology Collaboration in India.

The approval techniques follow two methods, namely automatic approval and government approval

Government Approval

The manufacturing and products should be compliant with the small scale industries Proposals which involve previous trademark agreement, joint ventures, technology transfer, etc In case of an extension of the foreign technology collaboration agreements which had been automatically approved earlier

3.19.3 Procedures for Approvals of Foreign Technology Agreements

The **Procedures for Approvals of Foreign Technology Agreements** were presented by the Central Government in the Statement on Industrial Policy in 1991. The Sec 39 C contains the clauses of the methods for better functioning of the Foreign Technology Agreements.

Approvals of Foreign Technology Agreements-At a glance

Procedures for Approvals of Foreign Technology Agreements helps in smooth settlement of the transfer of technical know how. With the introduction of the 1991 Economic Policy, foreign investments started inflowing in large amounts.

The transfer of foreign technology plays an important role in the Foreign Direct Investments. The Central government has allowed the transfer of technology in various industrial sectors. The economic policies elucidated the foreign exchange regulatory approvals for transfer of technology.

The Central Government presented the Statement on Industrial Policy in 1991, which provided simplified methods for the better governance of the Foreign Technology Agreements. The section dealing with the Foreign Technology Agreements is Sec 39C

Procedures for Approvals of Foreign Technology Agreements-Sec 39C

- The high priority industries would be given automatic approval under the norms of RBI, for foreign technology agreements, subject to a maximum limit of payments up to Rs. 1 crore.
- The royalty to be paid is restricted to 5 % in case of domestic sales, 8 % in case of exports and total payment should be 8 % on sales for a period of 10 years
- The royalty period should not exceed 7 years from the date of starting of the business or 10 years from the date mentioned in the agreement
- The royalty rates would be calculated in accordance with the standardized methods
- Industries apart from high priority ones would be allowed by the means of automatic approval in case no free foreign exchange is required in case of payments.
- Any other kinds of proposals would require particular approval under the general procedures
- Permissions pertaining to foreign testing of developed technological applications, employing foreign technicians
- The manufacturing and products should be compliant with the small scale industries

- In case of an extension of the foreign technology collaboration agreements which had been automatically approved earlier.

3.19.4 Standard Conditions Attached to Approval for Foreign Technology Agreements

The Standard Conditions Attached to Approval for Foreign Technology Agreements are one of the major tools for analyzing the agreement pertaining to foreign transfer of technology.

Standard Conditions Attached to Approval for Foreign Technology Agreements helps in effective inflow of foreign direct investments in India. The criteria for the approval of the Foreign Technology Agreements are-

Standard Conditions Attached to Approval for Foreign Technology Agreement

- The NRI holdings should not exceed the amount mentioned in the approval letter.
- The calculation of the royalty would be done, based on the net ex-factory sale price of the product not including the excise duties and subtracting the cost of standard bought-out components and the cost of imported components, including insurance, ocean freight, customs duties, etc.
- The royalty payment is contained to the licensed capacity and 25% excess pertaining to products needing industrial licensing or on units mentioned in the approval letter.
- For products not requiring industrial license, the restrictions are not applicable.
- A prior approval from the Central Government is necessary for the payment of royalty in case, the production crosses the quantity mentioned in the agreements.
- In case the orders are not been carried out during the time period of the agreement, the royalty is not payable beyond that period.
- The least guaranteed royalty is not allowed.
- Lump sum are to be paid by means of three equated installments such as the first one would be paid subsequent to the approval of agreement proposal by the Reserve Bank of India and it is registered to the foreign exchange, the second installment is given after the legal transfer of the documents pertaining to the technological know how, and the final installment is to be given after the production commences or after 4 years from the time of the approval.
- The exchange rates of the date of payments are followed in case of any payments.

- For hiring of foreign technicians, no permit is required, even if it is not mentioned in the approved agreement.
- In case of sending Indian personnel abroad for training, the company should approach the RBI.

The payments in case of foreign testing of basic products and materials are authorized by RBI.

3.20 Public sector policy

The public sector has been central to our philosophy of development. In the pursuit of our development objectives, public ownership and control in critical sector of the economy have played an important role in preventing the concentration of economic power, reducing regional disparities and ensuring that planned development serves the common good.

The Industrial Policy Resolution of 1956 gave the public sector a strategic role in the economy. Massive investments have been made over the past four decades to build a public sector, which has a commanding role in the economy. Today key selectors of the economy are dominated by the mature public enterprises that have successfully expanded production, opened up new areas of technology and built up a reserve of technical competence in a number of areas.

After the initial exuberance of the public sector entering new areas of industrial and technical competence, a number of problems have begun to manifest themselves in many of the public enterprises. Serious problems are observed in the insufficient growth in productivity, poor project management, over-manning, lack of continuous technological upgradation and inadequate attention to R&D and human resources development. In addition, public enterprises have shown a very Low rate of return on the capital invested. This has inhibited their ability to regenerate themselves in terms of new investments as well as in technology development. The result is that many of the public enterprises have become a burden rather than being an asset to the Government. The original concept of the public sector has also undergone considerable dilution. The most striking example is the takeover of sick units from the private sector. This category of public sector units accounts for almost one third of the total losses of central public enterprises. Another category of public enterprises, which does not fit into the original idea of the public sector being at the commanding heights of the economy, is the plethora of public enterprises which are in the consumer goods and services sectors.

It is time, therefore, that the Government adopt a new approach to public enterprises. There must be a greater commitment to the support of public enterprises, which are essential for the operation of the industrial economy. Measures must be taken to make these enterprises more growth oriented and technically dynamic. Units, which may be faltering at present but are potentially viable must be restructured and given a new lease of Life. The priority areas for growth of public enterprises in the future will be the following:

- Essential infrastructure goods and services.
- Exploration and exploitation of oil and mineral resources.
- Technology development and building of manufacturing capabilities in areas which are crucial in the long term development of the economy and where private sector investment is inadequate.
- Manufacture of products where strategic considerations predominate such as defence equipment.
- At the same time the public sector will not be barred from entering areas not specifically reserved for it.

In view of these considerations, Government will review the existing portfolio of public investments with greater realism. This review will be in respect of industries based on low technology, small scale and non-strategic areas, inefficient and unproductive areas, areas with low or nil social considerations or public purpose, and areas where the private sector has developed sufficient expertise and resources.

Government will strengthen those public enterprises, which fall in the reserved areas of operation or are in high priority areas or are generating good or reasonable profits. Such enterprises will be provided a much greater degree of management autonomy through the system of memoranda of understanding. Competition will also be induced in these areas. Government holdings in the equity share capital of these enterprises will be disinvested in order to provide further market discipline to the performance of public enterprises. There are a large number of chronically sick public enterprises incurring heavy losses, operating in a competitive market and serve little or no public purpose. These need to be attended to. The country must be proud of the public sector that it owns and it must operate in the public interest.

3.21 Monopolies and Restrictive Trade Practices Act (MRTP Act)

The principal objectives sought to be achieved through the MRTP Act are as follows:

i) Prevention of concentration of economic power to the common detriment, control of monopolies, and

ii) Prohibition of monopolistic and restrictive and unfair trade practices.

The MRTP Act became effective in June 1970. With the emphasis placed on productivity in the Sixth Plan, major amendments to the MRTP Act were carried out in 1982 and 1984 in order to remove impediments to industrial growth and expansion. This process of change was given a new momentum in 1985 by an increase of threshold limit of assets.

With the growing complexity of industrial structure and the need for achieving economies of scale for ensuring higher productivity and competitive advantage in the international market, the interference of the Government through the MRTP Act in investment decisions of large companies has become deleterious in its effects on Indian industrial growth. The pre-entry scrutiny of investment decisions by so-called MRTF companies will no longer be required. Instead, emphasis will be on controlling and regulating monopolistic, restrictive and unfair trade practices rather than making it necessary for the monopoly houses to obtain prior approval of Central Government for expansion, establishment of new undertakings, merger, amalgamation and takeover and appointment of certain directors. The thrust of policy will be more on controlling unfair or restrictive business practices. The MRTP Act will be restructured by eliminating the legal requirements for prior governmental approval for expansion of present undertakings and establishment of new undertakings. The provisions relating to merger, amalgamation, and takeover will also be repealed. Similarly, the provisions regarding restrictions on acquisition of and transfer of shares will be appropriately incorporated in the Companies Act.

Simultaneously, provisions of the MRTP Act will be strengthened in order to enable the MRTP Commission to take appropriate action in respect of the monopolistic, restrictive and unfair trade practices. The newly empowered MRTP Commission will be encouraged to require investigation *suo mote* or on complaints received from individual consumers or classes of consumers.

3.22 Navaratna companies

In 1997, the Government of India, had identified Nine Public sector enterprises (PSES) as potential and prestigious having authority to compete as a global giant in the global market. The number of Navaratna was raised from Nine to Eighteen. The Government is also proposing to confer **"Maha-navaratna"** to companies like **Gail, ONGC, Oil India,** NTPC to name a few.

The term Navaratna means nine gems ornament or a talisman in old days in the courts of King Vikramaditya and the Mughal emperor Akbar, Navaratna were a group of nine extra ordinary men in their respective courts. In case of Public sector enterprises, company which enhances its economical and operational capacity and invest up to Rs 1000 crore or 15% of their net worth on a single project without seeking governments approvals, are offered with the status of Navaratna.

In list of 1997, the original Navaratna companies were

1. BHEL
2. BPCL
3. HPCL
4. IOC
5. IPCL
6. NTPC
7. ONGC
8. SAIL
9. VSNL

As per the available information on November 2009 the list of Navaratna companies are enlisted below:-:

1. Bharat Electronics Limited
2. Bharat Heavy Electricais Limited
3. Bharat Petroleum Corporation Limited
4. Coal India Limited
5. GAIL (India) Limited
6. Hindustan Aeronautics Limited
7. Hindustan Petroleum Corporation Limited
8. Indian Oil Corporation Limited

9. Mahanagar Telephone Nigam Limited
10 National Aluminium Company Limited
11. NMDC Limited
12. NTPC Limited
13. Oil and Natural Gas Corporation Limited
14. Power Finance Corporation Limited
15. Power Grid Corporation of India Limited
16. Rural Electrification Corporation Limited
17. Shipping Corporation of India Limited
18. SAIL

The government of India has created another category called Miniratna. Which is further classified is two categories:

1. **Miniratnas-Category I:-** It includes PSES that have made profit for Last three years or earned a net profit of Rs 30 crore. It includes following companies.

1. Airport Authority of India
2. Balmer Lawrie & Co. Limited
3. Bharat Dynamics Limited
4. BEML Limited
5. Bharat Sanchar Nigam Limited
6. Bongaigaon Refineries & Petrochemicals Limited
7. Central Warehousing Corporation
8. Central Coalfields Limited
9. Chennai Petroleum Corporation Limited
10. Cochin Shipyard Limited
11. Container Corporation of India Limited
12. Dredging Corporation of India Limited
13. Engineers India Limited
14. Garden Reach Shipbuilders & Engineers Limited
15. Goa Shipyard Limited
16. Hindustan Copper Limited
17. HLL Lifecare Limited (Formerly Hindustan Latex Limited)
18. Hindustan Newsprint Limited
19. Housing & Urban Development Corporation Limited
20. India Tourism Development Corporation Limited
21. Indian Railway Catering & Tourism Corporation Limited
22. IRCON International Limited
23. Kudremukh Iron Ore Company Limited

24. Mazagaon Docks Limited
25. Mahanadi Coalfields Limited
26. Mangalore Refinery & Petrochemicals Limited
27. MMTC Limited
28. MSTC Limited
29. National Fertilizers Limited
30. Neyveli Lignite Corporation
31. NHPC Limited
32. Northern Coalfields Limited
33. Numaligarh Refinery Limited
34. Oil India Limited
35. Rashtriya Chemicals & Fertilizers Limited
36. Rashtriya Ispat Nigam Limited
37. RITES Limited
38. Satluj Jal Vidyut Nigam Limited
39. South Eastern Coalfields Limited
40. State Trading Corporation of India Limited
41. Telecommunications Consultants (India) Limited
42. Airports Authority of India (AAI)
43. CSR Group

2. **Miniratnas-Category II:-** It includes PSEs that have made profit for Last three years or earned a positive net profit. It includes following companies.

1. Broad cast Engineering Consultants(I) Limited
2. Central Mine Planning and Design Institute Limited
3. Educational Consultants (I) Limited
4. Engineering Projects (I) Limited
5. Ferro Scrap Nigam Limited
6. HMT (International) Limited
7. HSCC (India) Limited
8. India Trade Promotion Organization
9. Indian Medicines Pharmaceuticals Corporation Limited
10. Manganese Ore India Limited
11. MECON Limited
12. National Film Development Corporation Limited
13. PEC Limited
14. Rajasthan Electronics & Instruments Limited
15. Water & Power Consultancy (India) Limited

Summary

- The Economic history of India begins with the Indus valley civilisation which flourished between 2800 B.C. and 1800 B.C.
- By 300B.C. Era of Mughal Empire started which united down many sub continents of India as a result of which the trade routes became more secure and reliable .
- Between the period of 1st and 15th century A.D. India was estimated to have the largest economy and emerged as one of the fastest growing large economies of world
- The major agrarian struggles that later paved the way for the agricultural development of rural India are :- telangana Peasant struggle, Patiala Muzara Movement, land reforms Acts , tenancy reforms, land ceiling and many more .
- In 1967, the industrial licensing policy inquiry committee was setup , which recommended that larger industrial houses should be given license only for setting up industry y in sore and heavy investment sectors, there by necessitating re -orientation of industrial licensing policy .
- In 1969, the monopolies and restrictive trade practices (MRTP) act was introduced to enable the government to effectively control concentration of economic power.
- In 1997, the government of India had identified Nine public sectors enterprises (PSEs) as potential and prestigious having authority to compete as a global giant in the global market.

Objective Type Questions

1. Archeological Research in cities like Mohenjo-Daro , Harappa, Lothal and Rakhigarhi belongs to which civilisation

    (a) Mughal Empire

    (b) Indus Valley civilisation

    (c) Maurayan dynasty

    (d) Gupta Empire

2. The Era of Liberalization started after

    (a) 1947

    (b) 1950

    (c) 1960

    (d) 1980

3. Which Empire/Civilisation flourished between 321 BC and 185 BC?

   (a) 1947 (b) 1950

   (c) 1960 (d) 1980

4. Bhoodan Movement was launched by-

   (a) Dr. Vikram Sarabhai (b) Mahatma Gandhi

   (c) Jadish Chandra Bose (d) Acharya Vinoba Bhave

5. Who is regarded as the father of Indian Space Research Programme?

   (a) Dr. Vikram Sarabhai (b) Mahatma Gandhi

   (c) Jadish Chandra Bose (d) Acharya Vinoba Bhave

6. Indian Space Research Organisation (ISRO) was founded on

   (a) 15th August 1969 (b) 15th August 1979

   (c) 26th January 1969 (d) 26th January 1969

7. Which empire flourished between 1526 and 1700 BC?

   (a) 1947 (b) 1950

   (c) 1960 (d) 1980

8. The Industrial Licensing Policy Committee was setup in -

   (a) 1965 (b) 1966

   (c) 1967 (d) 1968

Answer to objective Questions

1. b 2. d 3. b 4. d 5. a 6. a 7. a 8. c

Review Questions

1. Define Industrialisation.
2. Briefly Discuss the Development of Industrialisation in India.
3. Discuss various events takes place after independence of India.
4. Enlist the various impacts of industrialisation upon various aspects of society.
5. What are the developments in science after Independence in India?
6. What were the objectives behind implements of Industrial policy?
7. What were the strategies followed for implemented of Industrial policy?

8. Write notes on:

   (1) Industrial policy Resolution-1956

   (2) Industrial Policy Statement -1977

   (3) Industrial Policy Statement-1980

   (4) Industrial Policy Statement-1991

9. Enlist the negative list of industries units' ineligible from exemption/ concession.

10. Discuss the aim of Foreign Technology Collaboration in India and what are the procedures for approval of foreign technology agreement.

11. Write Short notes on:

    (a) MRTP Act

    (b) Public Sector Policy

    (c) Industrial Licensing Policy

12. Enlist the names of Industries to be called under category of Navaratna Companies.

Chapter 4

Industrial Relations

4.1 Introduction

Industrial relations constitute one of the important and serious problems that is characterised by frequent changes, Industrial unrest and conflict ideologies at National and International level. It is a dynamic concept which depends upon the social changes, economical changes and change in political set up of a country. This concept gives us an art of living in a society or Industry for the common cause of production, efficiency, human well being and industrial progress. It comprises of a network of institutions, such as trade unionism, collective bargaining, employers, the law and the state, which are bound together by a set of common values and aspirations.

4.2 Concept and scope of Industrial relations

There is no unanimity on the meaning and scope of Industrial relations since different terms such as Labour management relations, employer-employee relations, union-management relations, personnel relations, human relations are in use and are used synonymously. Industrial relations generally mean existing relationship between the management and workers of an industry.

Dale Yoders quotes that "It is a whole field of relationship that exists because of the necessary collaboration of men and women in the employment process of an industry".

Tead and Met calfe quotes that "Industrial relations are the composite result of the attitudes and approaches of employers and employee towards each other with regard to planning, supervision, direction and coordination of the activities of an organization with a minimum of human efforts and frictions with an animating spirit of cooperation and with proper regard for the geniune well being of all members of the organization".

Allan flander quotes that the Subject of Industrial relations deals with certain regulated or institutionalized relationships in industry, personal or in the Language of sociology, "un structured" relationships have their importance for management and workers, but they lie outside the scope of a system of industrial relations".

Professor Elegg quotes that Industrial relations in the broadest terms as encompassing the rules governing employment together with the ways in which the rules are made and changed and their interpretation and administration.

It is a highly dedicate and dynamic concept which is concerned on the relations maintained at different stages of the organisation by the diverse, complex and composite needs and aspirations and attitudes and approaches among the participants. These relations may be formal or informal.

In Industrial relations, there are four basic elements:

(i) The organisations of workers and managements

(ii) The state

(iii) The managements

(iv) The workers

Fundamentally, Industrial relation represents organised relationship between two organised parties representing employers and employees and this leads to the growth of professional management. The scope of Industrial relation is not merely confined to the professional management and Labour management but it is a comprehensive concept which involves the sum total of relationships that exists at various levels of the organisational structure and denotes all types of inter group and intra group relations within industry, both formal and informal which consists of a complex network of relations that arise out of functional interdependence between workers and management and between industrial organisations and society.

It is also a relative concept, as it flourishes and damps relative to the changes in the economic, social and political conditions prevailing in a society and the Laws made by the state to regulate them.

4.3 Approaches to Industrial Relations

Any problem in industrial relations has to be approached on a multi disciplinary basis, drawing from the contributions of the above disciplines. The cause of an Industrial dispute may be, by nature, economic, social, psychological or political or a combination of any of them. Some of the approaches involved in solving industrial relationship problems are mentioned below.

4.3.1 Dunlop's Approach

Prof. John. T. Dunlop of Harvard University in his book Industrial Relations system (1958), presented an analytical framework of industrial relations. This book presents a general theory of industrial relations and provides tools of analysis to interpret and gain understanding of the widest possible range of industrial relations facts and practices.

Dunlop's defines an industrial relation system in the following ways:

An industrial relation system at any one time in its development is regarded as comprised of certain actors, certain contexts, an ideology, which binds the industrial relation system together, and a body of rules created to govern the actors at the workplace and work community. There are three sets of independent variables: the actors, the contexts and the 'ideology' of the system.

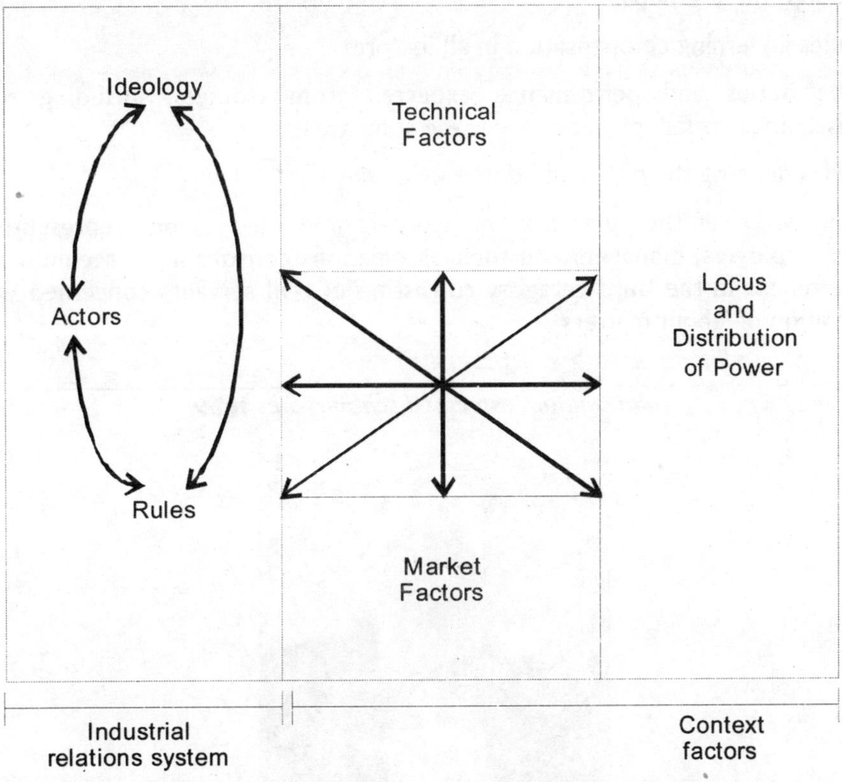

Fig. 4.1 The Dunlop's Industrial Relation system

The Actors in above Fig. 4.1 are:-

(i) Hierarchy of managers and their representatives in supervision.

(ii) Hierarchy of workers (non-managerial) and any spokesmen.

(iii) Specialised governmental agencies concerned with workers, enterprises and their relationships.

The first two actors are directly related with each other, but the third actor may or may not have any connection with the other two actors i.e. the role of specialised governmental agencies may be minor or constricted.

The ideology is a philosophy or set of ideas and beliefs commonly held by the actors that helps to bind or integrate the system together as an entity and it defines the role and place of each actor and the ideas, which each actor holds towards the place and function of the others in the system.

The Rules consists of procedure for establishing rules, the substantive rules and procedure for deciding their application to particular situation. These rules are classified into three groups:

1. Rules governing compensation in all its forms.
2. The duties and performance expected from workers, including rules of disciplines for failure to achieve these standards.
3. Rules defining the rights and duties of workers.

The actors who set the rules may be workers and their unions representing one category; employers, managers and their associations constituting a second category; and government in the third category consisting of civil servants concerned with the administration of labour matters.

John Dunlop, esteemed scholar, dies at 89

John Dunlop a labor negotiator and government official served as an adviser to many U.S presidents beginning with Franklin D. Roosevelt, Dunlop was secretary of labor under Gerald Ford, serving from March 1975 to January 1976.

Harvard President Emeritus Derek Bok, who co-authored the book "Labor and the American Community" (1970) with Dunlop, said of his colleague: "John Dunlop led a remarkable life. As a teacher, he helped develop generations of labor economists. As a scholar, he was a leading figure in furthering our understanding of labor markets and institutions. As a practitioner, he played an indispensable role in finding common ground between labor unions, employers, and government. While outwardly a thoroughly pragmatic problem-solver, he was at bottom a deeply principled man who always put the welfare of others above his own personal advancement. He had a profound influence on my life and leaves a void that can never be filled."

During his tenure at Harvard, Dunlop was a key figure in founding the Trade Union Program (now the Labor and Work Life Program); the Ph.D. Program In Business and Economics; the Joint Center for Housing Studies; and the Program in Business and Government. He was instrumental in founding and/or fundraising for many other academic programs and research centers.

He wrote numerous books and articles, including "Wage Determination Under Trade Unions" (1944); "Collective Bargaining: Principles and Cases" (1949); "Industrial Relations Systems" (1958); "Labor in the Twentieth Century" (ed., 1978); "Dispute Resolution, Negotiation and Consensus Building" (1984); and "A Stitch in Time: Lessons from the Apparel and Textile Industries" (with Abernathy, Hammond, and Weil, 1999).

4.3.2 The oxford approach

This approach says that "Industrial relations system is a study of institutions of job regulations. Flander considers every business organization as a social system of production and distribution, with a structured pattern of relationships. The "institution of job regulation was further categorised by him as internal and external, and it includes code of work rules, wage structure, internal procedure of joint consultation and grievance procedure.

This approach can be expressed as

G.R. = f (C.B.)

or G.R. = f (C.R.)

Where G.R. = Governing rules in Industrial relations

C.B. = Collective bargaining

C.R. = Conflict resolved through collective bargaining

4.3.3 Industrial Sociology approach

According to Industrial Sociologist, G. Margerism, the core of Industrial relations is the nature and development of the conflict itself i.e. conflict is the main cause which forms the base line of the study of Industrial relations.

According to this approach, there are two major conceptual levels of Industrial relations.

(a) **Intra-plant level:** In this level situational factors like job content, work task and technology and interaction factors produces three types of conflict-distributive., structural and human relations. These conflicts are being resolved through collective bargaining, structure analysis and man-management respectively.

(b) **Outside plant level:** This level deals with the situations outside the firm.

4.3.4 THE ACTION THEORY APPROACH

Like the systems model, the action theory approach takes the collective regulation of industrial labour as its focal point. The actors operate within a framework, which can at best be described as a coalition relationship. The actors, it is claimed, agree in principle to cooperate in the resolution of the conflict, their cooperation taking the form of bargaining. Thus, the action theory analysis of industrial relations focuses primarily on bargaining as a mechanism for the resolution of conflicts. Whereas the systems model of industrial relations constitutes a more or less comprehensive approach, it is hardly possible to speak of one uniform action theory concept.

4.3.5 THE MARXIST APPROACH

The Marxist approach is primarily oriented towards the historical development of the power relationship between capital and labour. It is also characterised by the struggle of these classes to consolidate and strengthen their respective positions with a view to exerting greater influence on each other. In this approach, industrial relations is equated with a power-struggle. The price payable for labour is determined by a confrontation between conflicting interests. The capitalist ownership of the enterprise endeavors to purchase labour at the lowest possible price in order to maximise their profits. The lower the price paid by the owner of the means of production for the labour he employs, the greater is his profit. The Marxist analysis of industrial relations, however, is not a comprehensive approach as it only takes into account the relations between capital and labour. It is rather, a general theory of society and of social change, which has implications for the analysis of industrial relations within what Marxists would describe as capitalist societies.

4.3.6 THE PLURALIST APPROACH

The focus is on the resolution of conflict rather than its generation, or, in the words of the pluralist, on 'the institutions of job regulation.' Kerr is one of the important exponents of pluralism. According to him, the social environment is an important factor in industrial conflicts. The isolated masses of workers are more strike-prone as compared to dispersed groups. When industrial jobs become more pleasant and employees' get more integrated into the wider society, strikes will become less frequent.

4.3.7 WEBER'S SOCIAL ACTION APPROACH

Weberian approach gives the theoretical and operational importance to "control" as well as to the power struggle to control work organisations – a power struggle in which all the actors in the industrial relations drama are caught up.

4.3.8 THE HUMAN RELATIONS APPROACH

The human relations approach highlights certain policies and techniques to improve employee morale, efficiency and job satisfaction. It encourages the small work group to exercise considerable control over its environment and in the process helps to remove a major irritant in labour-management relations.

4.3.9 THE GANDHIAN APPROACH

This approach to industrial relation is based upon fundamental principal of truth, non-violence and non-possession. This approach presumes the peaceful co-existence of

Mahatma Gandhi

Mohandas Karamchand Gandhi was born on **October 2**, 1869 in **Porbandar**, India.
He developed a method of action based upon the principles of courage, nonviolence and truth called Satyagraha. He believed that the way people behave is more important than what they achieve. Satyagraha promoted nonviolence and civil disobedience as the most appropriate methods for obtaining political and social goals. In 1915 Gandhi returned to India. Within 15 years he became the leader of the Indian nationalist movement.
Using the principles of Satyagraha he led the campaign for Indian independence from Britain.
More than once Gandhi used fasting to impress upon others the need to be nonviolent. India was granted **independence in 1947**, and partitioned into India and Pakistan. Gandhi had been an advocate for a united India where Hindus and Muslims lived together in peace.

capital and labour. Gandhi ji emphasized that if the employers follow the principal of trusteeship than there is no scope of conflict of interest between labour and management, Gandhi ji accepted the workers right to strike, but cautioned that they should exercise this right for a just cause and in a peaceful and non-violence manner and this method should only be resorted when all methods failed in getting employers response.

4.4 Features pertaining to Industrial Relations

- Industrial Relation do not emerge in vacuum they are born of employment relationship in an industrial setting. Without the existence of the two parties, i.e. labor and management, this relationship cannot exist. It is the industry, which provides the environment for industrial relations.
- Industrial Relation are characterized by both conflict and co-operations. This is the basis of adverse relationship. So the focus of Industrial Relations in on the study of the attitudes, relationships, practices and procedure developed by the contending parties to resolve or at least minimize conflicts.
- As the labor and management do not operate in isolations but are parts of large system, so the study of Industrial Relation also includes vital environment issues like technology of the workplace, country's socio-economic and political environment, nation's labor policy, attitude of trade unions workers and employers.
- Industrial Relation also involve the study of conditions conductive to the labor, managements co-operations as well as the practices and procedures required to elicit the desired co-operation from both the parties.
- Industrial Relations also study the laws, rules regulations agreements, awards of courts, customs and traditions, as well as policy framework laid down by the governments for eliciting co-operations between labor and management. Besides this, it makes an in-depth analysis of the interference patterns of the executive and judiciary in the regulations of labor–managements relations.

4.5 Industrial Relations Factors that influence the behavior of people

In fact, Industrial Relation encompasses all such factors that influence behavior of people at work. A few such important factors are details below:

1. **Institution:** It includes government, employers, trade unions, unions federations or associations, government bodies, labor courts, tribunals and other organizations which have direct or indirect impact on the industrial relations systems.

2. **Characters** : It aims to study the role of workers unions and employers' federations officials, shop stewards, industrial relations officers/ manager, mediator/conciliators / arbitrator, judges of labor court, tribunal etc.

3. **Methods** : Focus on collective bargaining, workers' participation in the Industrial Relation schemes, discipline procedure, grievance re-dressal machinery, dispute settlements machinery working of closed shops, union reorganization, organizations of protests through methods like revisions of existing rules, regulations, policies, procedures, hearing of labor courts, tribunals etc.

4. **Contents** : Includes matter pertaining to employment conditions like pay, hours of works, leave with wages, health, and safety disciplinary actions, lay-off, dismissals retirements etc., laws relating to such activities, regulations governing labor welfare, social security, industrial relations, issues concerning with workers' participation in management, collective bargaining, etc.

4.6 Objectives of Industrial Relation

To safeguard the interest of labor and management by securing the highest level of mutual understanding and good-will among all those sections in the industry which participate in the process of production.

To avoid industrial conflict or strife and develop harmonious relations, which are an essential factor in the productivity of workers and the industrial progress of a country.

To raise productivity to a high level in an era of full employment by lowering the tendency of high turnover and frequency absenteeism.

To establish and nurse the growth of an Industrial Democracy based on labor partnership in the sharing of profits and of managerial decisions, so that ban individuals personality may grow its full stature for the benefit of the industry and of the country as well.

To eliminate, as far as is possible and practicable, strikes, lockouts and gheraos by providing reasonable wages, improved living and working conditions, said fringe benefits.

To regulate government control of such plants and units as are running at a loss or in which productions has to be regulated in the public interest.

Improvements in the economic conditions of workers in the existing state of industrial managements and political government.

Control exercised by the state over industrial undertaking with a view to regulating production and promoting harmonious industrial relations.

Socializations or rationalization of industries by making he state itself a major employer

Vesting of a proprietary interest of the workers in the industries in which they are employed.

4.7 The main aspect of Industrial Relations are :-

- Labor Relations, i.e. relations between union and management.
- Employer-employees relations, i.e. relations between management and employees.
- Group relations, i.e. relations between various groups of workmen.
- Community or Public relations, i.e. relations between industry and society.
- Promotions and development of healthy labor-managements relations.
- Maintenance of industrial peace and avoidance of industrial strife
- Development of true industrial Democracy.

4.8 Effects of poor Industrial Relations

Poor Industrial Relation produces highly disquieting effects on the economic life of the country. We may enumerate the ill-effects of poor Industrial Relations as under:

1. **Multiplier effects:** Modern industry and for that matter modern economy are interdependent. Hence although the direct loss caused due to industrial conflict in any one plant may not be very great, the total loss caused due to its multipliers effect on the total economy is always very great.

2. **Fall in normal tempo:** poor Industrial Relations adversely effect the normal tempo of work so that work far below the optimum level. Costs build up. Absenteeism and labor turnover increase. Plants discipline breaks down and both the quality and quality of production suffer.

3. **Resistance of change:** Dynamic industrial situation calls for change more or less continuously. Methods have to be improved. Economics have to be introduced. New products have to be designed, produced and put in the market. Each of these tasks involves a whole chain of changes and this is resisted bitterly if these are industrial conflict.

4. **Frustration and social cost:** every man comes to the work place not only to earn a living. He wants to satisfy his social and egoistic needs also. When he finds difficulty in satisfying these needs he feels frustrated. Poor Industrial Relations take a heavy toll in terms of human frustration. They reduce cordiality and aggravate social tension.

4.9 Measures For Improving Industrial Relations

The following measures should be taken to achieve good industrial relations:

- **Strong and Stable Union:** A strong and stable union in each industrial enterprise is essential for good industrial relations which represents the majority of workers and negotiate with the management about the terms and conditions of service.
- **Mutual Trust:** Both management and labor should help in the development of an atmosphere of mutual cooperation, confidence and respect. Management should adopt a progressive outlook and should recognize the rights of workers. Similarly, labor unions should persuade their members to work for the common objectives of the organization. Both the management and the unions should have faith in collective bargaining and other peaceful methods of settling disputes.
- **Workers' Participation in Management:** The participation of workers in the management of the industrial unit should be encouraged by making effective use of works committees, joint consultation and other methods. This will improve communication between managers and workers, increase productivity and lead to greater effectiveness.
- **Mutual Accommodation.** The employers must recognize the right of collective bargaining of the trade unions. In any organization, there must be a great emphasis on mutual accommodation rather than conflict or uncompromising attitude. The approach must be of mutual "give and take rather than "take or leave." The management should be willing to co-operate rather than blackmail the workers.
- **Sincere Implementation of Agreements.** The management should sincerely implement the settlements reached with the trade unions. The agreements between the management and the unions should be enforced both in letter and spirit.
- **Sound Personnel Policies:** The following points should be noted regarding the personnel policies. The policies should be:
    - Formulated in consultation with the workers and their representatives if they are to be implemented effectively.
    - Clearly stated so that there is no confusion in the mind of anybody.
    - Implementation of the policies should be uniform throughout the organization to ensure fair treatment to each worker.
- **Government's Role:** The Government should play an active role for promoting industrial peace. It should make law for the compulsory recognition of a representative union in each industrial unit. For restoring industrial harmony Government should intervene to settle disputes if the management and the workers are unable to settle their disputes.
- **Progressive Outlook:** There should be progressive outlook of the management of each industrial enterprise. It should be conscious of its obligations and responsibilities to the owners of the business, the employees, the consumers and

the nation. The management must recognize the rights of workers to organize unions to protect their economic and social interests.

4.10 Importance of Industrial Relations

The healthy industrial relations are keys to the progress. Their significance may be discussed as under -

1. **Uninterrupted production** – The most important benefit of industrial relations is that this ensures continuity of production. This means, continuous employment for all from manager to workers. The resources are fully utilized, resulting in the maximum possible production. There is uninterrupted flow of income for all. Smooth running of an industry is of vital importance for several other industries; to other industries if the products are intermediaries or inputs; to exporters if these are export goods; to consumers and workers, if these are goods of mass consumption.

2. **Reduction in Industrial Disputes** – Good industrial relation reduce the industrial disputes. Disputes are reflections of the failure of basic human urges or motivations to secure adequate satisfaction or expression which are fully cured by good industrial relations. Strikes, lockouts, go-slow tactics, gherao and grievances are some of the reflections of industrial unrest which do not spring up in an atmosphere of industrial peace. It helps promoting co-operation and increasing production.

3. **High morale** – Good industrial relations improve the morale of the employees. Employees work with great zeal with the feeling in mind that the interest of employer and employees is one and the same, i.e. to increase production. Every worker feels that he is a co-owner of the gains in industry. The employer in his turn must realize that the gains in industry are not for him alone but they should be shared equally and generously with his workers. In other words, complete unity of thought and action is the main achievement of industrial peace. It increases the place of workers in the society and their ego is satisfied. It naturally affects production because mighty co-operative efforts alone can produce great results.

4. **Mental Revolution** – The main object of industrial relation is a complete mental revolution of workers and employees. The industrial peace lies ultimately in a transformed outlook on the part of both. It is the business of leadership in the ranks of workers, employees and Government to work out a new relationship in consonance with a spirit of true democracy. Both should think themselves as partners of the industry and the role of workers in such a partnership should be recognized. On the other hand, workers must recognize employer's authority. It will naturally have impact on production because they recognize the interest of each other.

5. **New Programmes** – New programmes for workers development are introduced in an atmosphere of peace such as training facilities, labor welfare facilities etc. It increases the efficiency of workers resulting in higher and better production at lower costs.

6. **Reduced Wastage** – Good industrial relations are maintained on the basis of cooperation and recognition of each other. It will help increase production. Wastages of man, material and machines are reduced to the minimum and thus national interest is protected.

Thus, from the above discussion, it is evident that good industrial relation is the basis of higher production with minimum cost and higher profits. It also results in increased efficiency of workers. New and new projects may be introduced for the welfare of the workers and to promote the morale of the people at work.

An economy organized for planned production and distribution, aiming at the realization of social justice and welfare of the massage can function effectively only in an atmosphere of industrial peace. If the twin objectives of rapid national development and increased social justice are to be achieved, there must be harmonious relationship between management and labor.

4.11 Difference Between Industrial Relations and Human Relations

The term "Industrial Relations" is different from "Human Relations". Industrial relations refer to the relations between the employees and the employer in an industry. Human relations refer to a personnel-management policy to be adopted in industrial organizations to develop a sense of belongingness in the workers which in turn improves their efficiency and treat them as human beings and make a partner in industry.

Industrial relations cover the matters regulated by law or by collective agreement between employees and employers. On the other hand, problems of human relations are personal in character and are related to the behavior of worker where morale and social elements predominated. Human relations approach is personnel philosophy which can be applied by the management of an undertaking. The problem of industrial relations is usually dealt with a three levels – the level of undertaking, the industry and at the national level. To sum up the term "Industrial Relations" is more wide and comprehensive and the term "Human Relations" is a part of it.

4.12 Determining Factors of Industrial Relations

Good industrial relations depend on a great variety of factors. Some of the more obvious ones are listed below:

1. **History of industrial relations** – No enterprise can escape its good and bad history of industrial relations. A good history is marked by harmonious

relationship between management and workers. A bad history by contrast is characterized by militant strikes and lockouts. Both types of history have a tendency to perpetuate themselves. Once militancy is established as a mode of operations there is a tendency for militancy to continue. Or once harmonious relationship is established there is a tendency for harmony to continue.

2. **Economic satisfaction of workers** – Psychologists recognize that human needs have a certain priority. Need number one is the basic survival need. Much of men conducted are dominated by this need. Man works because he wants to survive. This is all the more for underdeveloped countries where workers are still living under subsistence conditions. Hence economic satisfaction of workers is another important prerequisite for good industrial relations.

3. **Social and Psychological satisfaction** – Identifying the social and psychological urges of workers is a very important steps in the direction of building good industrial relations. A man does not live by bread alone. He has several other needs besides his physical needs which should also be given due attention by the employer. An organization is a joint venture involving a climate of human and social relationships wherein each participant feels that he is fulfilling his needs and contributing to the needs of others. This supportive climate requires economic rewards as well as social and psychological rewards such as workers' participation in management, job enrichment, suggestion schemes, re-dressal of grievances etc.

4. **Off-the-Job Conditions** – An employer employs a whole person rather than certain separate characteristics. A person's traits are all part of one system making up a whole man. His home life is not separable from his work life and his emotional condition is not separate from his physical condition. Hence for good industrial relations it is not enough that the worker's factory life alone should be taken care of his off-the-job conditions should also be improved to make the industrial relations better.

5. **Enlightened Trade Unions** – The most important condition necessary for good industrial relations is a strong and enlightened labor movement which may help to promote the status of labor without harming the interests of management, Unions should talk of employee contribution and responsibility. Unions should exhort workers to produce more, persuade management to pay more, mobilize public opinion on vital labor issues and help Government to enact progressive labor laws.

6. **Negotiating skills and attitudes of management and workers** – Both management and workers' representation in the area of industrial relations come from a great variety of backgrounds in terms of training, education, experience and attitudes. These varying backgrounds play a major role in shaping the character of industrial relations. Generally speaking, well-trained and experienced negotiators who are motivated by a desire for industrial peace

create a bargaining atmosphere conducive to the writing of a just and equitable collective agreement. On the other hand, ignorant, inexperienced and ill-trained persons fail because they do not recognize that collective bargaining is a difficult human activity which deals as much in the emotions of people as in their economic interests. It requires careful preparation and top –notch executive competence. It is not usually accomplished by some easy trick or gimmick. Parties must have trust and confidence in each other. They must possess empathy, i.e. they should be able to perceive a problem from the opposite angle with an open mind. They should put themselves in the shoes of the other party and then diagnose the problem. Other factors which help to create mutual trust are respect for the law and breadth of the vision. Both parties should show full respect for legal and voluntary obligations and should avoid the tendencv to make a mountain of a mole hill.

7. **Public policy and legislation**: - when Government, regulates employee relations, it becomes a third major force determining industrial relations the first two being the employer and the union. Human behavior is then further complicated as all three forces interact in a single employee relation situation. Nonetheless, government in all countries intervenes in management – union relationship by enforcing labor laws and by insisting that the goals of whole society shall take precedence over those of either of the parties. Government intervention helps in three different ways 1) it helps in catching and solving problems before they become serious. Almost every one agrees that it is better to prevent fires them to try stopping them after they start; 2) It provides a formalized means to the workers and employers to give emotional release to their dissatisfaction; and 3) It acts as a check and balance upon arbitrary and capricious management action.

8. **Better education**: - with rising skills and education workers' expectations in respect of rewards increase. It is a common knowledge that the industrial worker in India is generally illiterate and is misled by outside trade union leaders who have their own axe to grind. Better workers' education can be a solution to this problem. This alone can provide worker with a proper sense of responsibility, which they owe to the organization in particular, and to the community in general.

9. **Nature of industry**: - In those industries where the costs constitute a major proportion of the total cast, lowering down the labor costs become important when the product is not a necessity and therefore, there is a little possibility to pass additional costs on to consumer. Such periods, level of employment and wages rise in decline in employment and wages. This makes workers unhappy and destroys good industrial relations.

4.13 Industrial Relations Programme

Today's professional industrial relations director, or by whatever title he is designated, no longer views his job as personalizing management, or that of a social worker in a factory, or a union buster, he looks upon his department as an adjunct to management supervision at all levels; he keeps other executives informed about new discoveries, programme trends and needs. At the same time, he provides efficient service in the operation of several centralized services.

A successful industrial relations programme reflects the personnel viewpoint, which is influenced by three main considerations:

a) Individual thinking

b) Policy awareness, and

c) Expected group reaction

Individualized thinking makes if imperative for the administrator to consider the entire situation in which the affected individual is placed. Policy awareness underscores the idea of the consistency of treatment and the precedent value of any decision which a management takes; while expected group reaction balances what we know of human nature in groups against an individual's situation in the light of the policy that has been formulated and implemented. In all these different circumstances, reality demands that all the three aspects of the personnel viewpoint should be considered at once in terms of the past, the present and the future. This viewpoint is held at all the levels of management from the top to the bottom, from the top executives and staff to the line and supervisory personnel.

4.14 Scope of Industrial Relations Work

The staff employed in the industrial relations department should know the limitations within which he has to function. The industrial relations director generally has several assistants who help him to perform his functions effectively, and he usually reports directly to the president or chairman of the board of directors of an organization.

The functions of the industrial relations staff are -

- Administration, including overall organization, supervision and co-ordination of industrial relations policies and programmes.
- Liaison with outside groups and personnel departments as well as with various cadres of the management staff.
- The drafting of regulations, rules, laws or orders and their construction and interpretation.
- Position classification, including overall direction of job analysis, salary and wage administration, wage survey and pay schedules.

- Recruitment and employment of workers and other staff.
- Employment testing, including intelligence tests, mechanical aptitude tests and achievement tests.
- Placement, including induction and assignment.
- Training of apprentices, production workers, foremen and executives.
- Employee counseling on all types of personnel problems-educational, vocational, health or behavior problems.
- Medical and health services.
- Safety services, including first aid training.
- Group activities, including group health insurance, housing, cafeteria programmes and social clubs.
- Suggestion plans and their uses in labor, management and production committees.
- Employee relations, specially collective bargaining with representatives and settling grievances.
- Public relations.
- Research in occupational trends and employee attitudes, and analysis of labor turnover.
- Employee records for all purposes.
- Control of operation surveys, fiscal research and analysis.
- Benefit, retirement and pension programmes.

4.15 Functional Requirements of a Successful Industrial Relations Programme

The basic requirements on which a successful industrial relations programme is based are :-

a) **Top Management Support:** - Since industrial relations is a functional staff service, it must necessarily derive its authority from the line organization. This is ensured by providing that the industrial relations director should report to a top line authority to the president, chairman or vice president of an organization.

b) **Sound Personnel Policies:** - These constitute the business philosophy of an organization and guide it in arriving at its human relations decisions. The purpose of such policies is to decide, before any emergency arises, what shall be done about the large number of problems which crop up every day during the working of an organization. Policies can be successful only when they are followed at all the level of an enterprise, from top to bottom.

c) **Adequate Practices should be developed by professionals:** - A system of procedures is essential if intention is to be properly translated into action. The procedures and practices of an industrial relations department are the "tool of management" which enables a supervisor to keep ahead of his job that of the time-keeper, rate adjuster, grievance reporter and merit rater.

d) **Detailed Supervisory Training** :- To ensure the organizational policies and practices are properly implemented and carried into effect by the industrial relations staff, job supervisors should be trained thoroughly, so that they may convey to the employees the significance of those policies and practices. They should, moreover, be trained in leadership and in communications.

e) **Follow-up of Results:** - A constant review of an industrial relations programme is essential, so that existing practices may be properly evaluated and a check may be exercised on certain undesirable tendencies, should they manifest themselves. A follow up of turnover, absenteeism, departmental morale, employee grievances and suggestion; wage administration, etc. should be supplemented by continuous research to ensure that the policies that have been pursued are best fitted to company needs and employee satisfaction. Hints of problem areas may be found in exit interviews, in trade union demands and in management meetings, as well as in formal social sciences research.

4.16 Future Strategies of Industrial Relations in India

Healthy IR in business and Industry are of paramount importance for increasing productivity. Industrial Law has played a crucial role in shaping Industrial Relations in this country. The Labor Laws seek to protect the interests of weaker segment in Industrial Relations and it has all along been the labour. The era of protective labour laws began with the enactment of Factories Act, which is still continuing and literally it has been taken over by our Hon'ble Judicial System.

Periphery of Industrial Law and concept of social justice are expanding. By a process of liberal and beneficial interpretation, the judiciaries have sought to expand the horizon of social justice with the result that no area relating to Industrial Relations is immune from judicial intervention. There is no other country in the world perhaps with the possible exception of West Germany, where there is so much job security as in India.

I would not like to go in to the journey of courts past judgments from where reinstatement of workers, who participated in an illegal strike to validation of threatening and abusive language by a worker to his superior took place, but it will be imperative to mention here the sharp change in judicial thinking in recent past. No doubt It has come through after long time but judicial system has underlined the need of discipline and efficiency to improve productivity and come heavily on errant workers. In one judgment S.C. has observed: "This Court has come a long way from its earlier view points. The recent trend in the decisions of this Court seek to strike a balance between the earlier approach of the industrial relation wherein only the interest of the workmen was sought to be protected with the avowed object of fast industrial growth of the country. In several decisions of this Court it has been noticed that how discipline at the workplaces/industrial undertaking received a set back. In view of the change in economic policy of the country, it may not now be proper to allow the employees to break the discipline with impunity. Our country is governed by rule of law. All actions, therefore, must be taken in accordance with law. Law declared by this Court in terms of Article 141 of the Constitution of India, as noticed in the decisions, categorically demonstrates that the Tribunal would not normally interfere with the quantum of punishment imposed by the employers unless an appropriate case is made out therefore. The Tribunal being inferior to that of this court was bound to follow the decisions of this Court. The Tribunal can neither ignore the ratio laid down by this Court not refuse to follow the same."

While one can appreciate the anxiety of our Government to improve the lot of the workman by protective legislation, too much protection is obviously undesirable. Most of the laws are one sided and operate to the detriment of employer and the industry and the result is growing indiscipline, corruption and loss of efficiency and productivity among labour force. I feel employer do deserve some quota of social justice, but there is nothing to protect their interests. Sorry state of affairs remain that despite the existence of protective and beneficial legislation, we have not been able to secure harmonious industrial relations. I must say that out labour legislation has failed to tackle the situation. We should not think of Industrial prosperity, harmonious IR and higher efficiency through legislation.

VRS is the brain child of such a situation created by over protected legislations of India. Simultaneously, it has every reason to effect the Industrial Relations science. Some HRD Chiefs believe that once the VRS is implemented in the Organisation to keep its survival, they will be able to knit the remaining smaller number of people in to a cohesive team. But in practice it never happens. VRS creates more IR problems some times. Unknowingly almost surreptitiously, all the remaining people become extremely suspicious about management intention, the entire top management in their eyes losses the credibility. They carry the nagging feeling that if this company can do it to so many people, who have worked so long, then why not for them also tomorrow and in this process, many good performers start looking for new openings. So HR Manager has to plan post VRS strategies simultaneously to avoid any damage at IR front. VRS is a double edged weapon, which if used negligently, can cut the performer even. We are fastly travelling in 21st Century. A century of knowledge management and technology. To my mind, better understanding, negotiation in good faith, productivity related wage increase, fixed team appointments, proactive approach of workers and unions, better work environment and work culture would be new dimensions of good IR in 21st Century.

Before we further move forward, one has to remember that the coming time will mainly centre around customer, cost, competition and care for MAN, Therefore, we may anticipate what sort of spectrum may emerge in future for Indian industries in the changing environment. We should remember that protected economy is the song of past days and free economy with divergent forces has started making its feel at this stage. And as consequence of it, certain things are likely to happen and let us glance at them.

The population of the country is now more than 100 crores. Unemployment which is acute today will further be aggravated leading to a large unskilled population who can not be employed in industries which may need more and more qualified, skilled and technical hands. Automation, mechanization, robotisation etc., will greatly reduce demand for unskilled while creating short supply of trained and experienced hands. MNCs and big business houses and corporate bodies will compete acquiring the cream of the country while a large number will be flying out of the country for better career opportunities abroad. Downsizing the industries and gradual reduction in industrial employment even seen in the recent past can lead to grave unemployment problem and labour unrest adversely affecting industrial growth and economy of the country. But worker profile will be different who will require all care and attention to work in competitive and challenging environment.

Future perceptions and strategies of Industrial Relations

With the changes, the economic, cultural scenario of the Century is going to be radically different. The Changing industrial climate in terms of vision, value and mission will dictate the norms of business ethics.

There are changes in the nature of labour force, union approaches to management process, physical and behaviour characteristics of internal forces, all affecting the strategies of Organisation.

But at the same time, we must face the reality that industrial relations in the law is a function of balance of power between collective groups. Whereas sound HRM practices should, and will, prevail at the level of individual employee relationship, the force of collectivism-that great emasculator of individual personality-has to be reckoned with. One of the harsh reality in such a situation is the wall that has become thicker and higher and stronger between the front-line manager supervisor and the individual employee. This process has to be stopped so that it may not create not only a communication gap and a feeling of estrangement and mistrust between the two most vital and productive partners at work.

I feel that earlier the managers pick up courage and organised employees become more resilient on this issue, the better will it be for the future of our industrial relations and progress. It is time we arrive at a consensus that whereas on collective issues the unions have a codetermined role as equal partners, we should leave individual relationship and his development at workplace between the employee and his frontline manager.

The management of IR will be part of the overall Organisation strategies. Line manager will have to increasingly manage the industrial relations at the shop-floor and work places. Negotiating with the unions-which essentially is an exercise in putting a cross management's view point-will have to be a team effort wherein knowledge and expertise from all disciplines will have to be marshaled and presented in a unified manner. Managers will have to speak with one voice and be ready to carry conviction where it is most difficult but most productive also. If we have to eliminate the unfortunate we-they situation that prevails between HR and line management, the interaction will have to come from the top management team so that sound HR Management practices are seen as a total organisational commitment to the enhancement of human excellence.

Excellent industrial relations are and will continue to be the end-product of sound HR policies and programmes based on human engineering, so HR managers have to ensure that their policies are transparent, based on performance thereby enhancing human values. Employee participation, Group dynamics leading to team spirit, both blue and white collar motivated employees and communication at last as nervous system of Organisation will form the core dimensions of excellent IR.

Our HR managers at all levels will have to earn "credibility" as professionals par excellence. They must humanise industry in their own limited sphere of professionalisation. Let them open a new chapter of Human Resource Management (HRM) by knocking out the legalistic framework, Let them abhor Courts of Law and all that legal entanglements connote. Let them have this mindset, this philosophy of man-management, and this sense of directness in HR management as 'professionals' in pursuit of excellence. They have to be the human engineers in the final analysis.

Employee Synergy:

The present dimensions of different cadres of employees as workers and managers will gradually diminish. The growing awareness of roles and responsibilities and the increasing levels of education will convert the less informed workers into knowledge workers who in turn will comfortably compete with their more privileged superiors-the Managers. The distinction thus will disappear, so traditional approach of maintaining IR will not work. HR Manager has to be fully prepared, competent, transparent and logical while dealing with workforce.

Consolidation of bargaining strengths

The bargaining strengths of Trade Unions have brought remarkable dividends to the working class all over the world. The managements have not recognised this fact that the less united group of employees, namely, the officers and staff have not been treated fairly and equitable. This denial of justice has been playing heavily on the white collared personnel and the formation of the federations in Banks and Public Sector Undertakings has been a pointer in this direction. Realising that the only way to receive justice from the hands of the management is to force them to understand that they should not be treated in a casual manner. This objective will be sought to be satisfied through consolidation of their strength in the form of associations. To meet out this situation, HR manager has to equip not only himself but complete team of management through interpersonal skill development and enhancement of competence level.

This situation may be diffused by taking care of middle managers in terms of improving their quality of life and paying compensation well so that they may not divert to formation of formal associations.

Information technology is bound to create an impact on industrial relations as both the parties - management as well as workers will use this tool to strengthen their bargaining power. It will be difficult rather impossible for HR managers to hide any business information from workers & union to take undue advantage while bargaining. Workers will also be ready with full facts and figures in their favour to justify their demands as management use to have. So HR managers have to be more careful and develop new negotiation skills in this regard.

Internal Leadership

External leadership has largely been replaced in many developed and developing countries with the advent of liberalised and enlightened outlook on the part of employers and a sense of commitment and concern in the employees. This will brighten the prospects of nurturing internal leadership among the employees. The frequent confrontations resulting in agitations, stoppages of work and closures of Industrial establishments will become a nightmare for all those who are concerned with productivity improvements and organisational excellence.

Empowerment

Employees should be empowered to acquire constructive roles in the management of the resources. When they will have a better say in the matter which govern the process of utilisation of resources, chances of conflicts and confrontation will be minimal.

Compensation Strategy

Competitive compensation and remuneration system with thrusts for 'payment by results will demand greater focus in the future. These will include Incentive Schemes, Variable Performance based pay, Annual Rewards and Bonus Schemes, Employee Stock Option Schemes etc, which will need to be related to and support the organisational objectives. At the same time, the control of unionised employee cost will continue to demand a lot of effort-this does not imply low salaries, but high productivity and manpower utilisation with competitive payments. Review and redesign of Retirement Benefits and Medical Assistance/Insurance during the service period, and post retirement benefits, will also be necessary. Retention and motivation of key employees will remain a major function and challenge for which planned IR strategy will be important.'

Few more strategies

- ***Proactive, system wide interventions with emphasis on fit:***

Linking HRM with strategic planning, cultural changes, pre-screening placement, promotions, lay-offs based on how individuals will fit with desired new culture, co-coordinating people and structural changes.

- ***People are social capital capable of development:***

Increasing power of development HRM functions, forming work-groups, using skill-based pay to broaden employees' competencies, heavy emphasis on training, more open communications, greater participation to develop people's commitment, more concern about employment security.

- ***Coincidence of interests between stakeholders can be developed:***

Joint union-management efforts based on improving quality of work life and organisational effectiveness, emphasis on employees self-esteem with positive human and business outcomes. Open channels of communications to build trust, commitment of greater sharing of information of business with unions, work groups, organisational units greater communication about design and administration of pay systems.

To meet these strategic role expectations, the human resources manager in the organisation will need specific knowledge, skills and behavior.

People are key to success. This is more relevant today and in time to come in ever changing environment and market economy in which industry functions. Leadership is a relationship with employees.

HR Managers have to influence the behavior of employees and develop skill, knowledge, capability to cope with the changing needs through leadership. He is to lead the people as individuals and members of teams. He has to be flexible and "look outside the box" to innovate, create, interpret and motivate people to work for mutual benefit. He cannot remain as a "tool box", but must be able to think and make people think and change to work in competitive environment. If necessary he has to initiate actions and lead to change law relating to company, labour, money market etc., necessary for industrial growth in the country.

It goes without saying that the effective management of Industrial Relations will continue to demand high priority in the coming era, not just to 'do well', but also for the survival of both, the organisation, as well as the employees, jobs. The HR Managers will have to develop more effective strategies to build understanding and co-operation for this common objective, before being too late. Instead of confronting, partnership attitudes between unions, employees and management will have to be created.

Maturity from all will need considerable changes in management styles, impartial attitude building exercises and development work spearheaded by the HR managers. Productivity improvement, managing flexibility, rapid adaption of technological up gradation and fast response to market requirements will also warrant the HR Managers to manage such changes with minimum disruption-an undoubting task.

And the final analysis

For maintaining, good IR, in future, responsibility should be shifted from HR Manager to line manager. It has to start from selection. People should be raw talent, Expectations must be set, organisation has from these people, not by setting out the steps they must take, but by communicating the end result that they must ensure. Individuals have to be strengthened and putting them at right fit for their talent. HR Manager should train line manager for this new role and arm them with HR tools so they could contribute effectively in keeping good IR and develop HR. And above all, you need to create a better workplace, so that employees are highly engaged in their work. Engaged employees treat organisation's concerns as their own, because they identify them closely with the company. The result is higher productivity better reward for people and of course good IR and HRD.

4.17 Government's role in Industrial relation

After Independance, In India, the government has been playing a comprehensive and vital in shaping out the structured pattern of Industrial. The government evolves through tripartite agreement which is a guide lines for shaping employes-employee relations. It accepts the responsibility of ensuring conformity to these norms through the administrative and judicial mechanism. It enacts legislation on labour and implements both the substantive and procedural Laws. The industrial relations policy of the government forms parts of the broader labour policy. This is stated by central industrial relations machinery (CIRM)

Central industrial relations machinery (CIRM)

This organisation is controlled by chief labour commissioner and performs actions like enforcement of labour Laws, Maintaining Industrial relation's and trade union membership verification. This organisation consists of 25 officers at Head Quarters and 253 officers in the field and has its offices spreaded all over the country.

Functions of CRIM

- Prevention and settlement of Industrial disputes
- Enforcement of labour Laws
- Implementation of awards
- Quasi-Judicial functions like inspection, controlling etc.

- Verification of the Trade union membership.
- Welfare of labour employed in industries.
- Investigations of complaints according to the rules of Ministry of labour and employment.
- Assisting and advising Ministry of Labor and employment in case of conflicts and training of central Labor service officers.
- Conducts meeting between Minimum wages advisory Board and notifying variable Dearness allowance at every six months.
- Assists central advisory contract Labor Board in order to examine the prohibition of contract Labour in different units or industries.

Major industrial Relations events during the year 2004-2005 in which CIRM Played important role are mentioned below:

1. Flash strike in Air India

A flash strike in Air India was reported to CLC(C) by Shri Bhargav, director (HRD), Air India. On 21st May, 2004 at 9 pm. strike was called by Air India Employees Guild against verification of membership of the employees of Air India (Ground category) undertaken by CLC(C) organisation, being advised, by Ministry of Labour and Employment on the request of Ministry of Civil Aviation. CLC(C), immediately contacted Union leaders over phone and requested them to withdraw flash strike and invited for discussions. On his request, union agreed to call off the strike and attended discussion with CLC(C) on 24.05.2004.

2. Syndicate Bank Strike averted

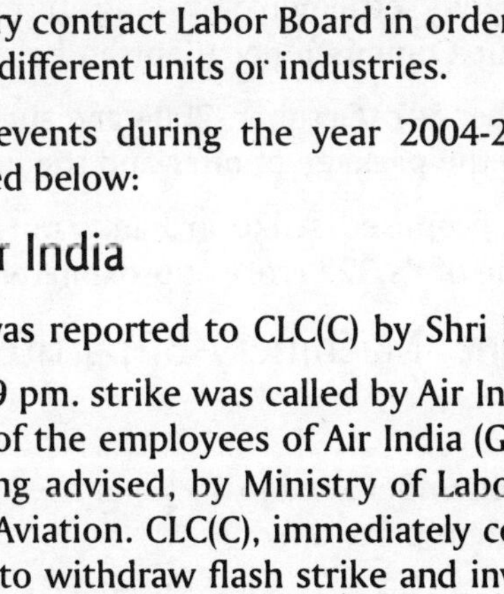

Unions operating in Syndicate Bank Served notice to go on Strike on 4.06.2004 in protest against the decision of the Bank to abolish the Special Assistance Cadre. The matter was seized into conciliation by the RLC(C) Bangalore and after prolonged discussions and persuasion, management of the bank and union leaders signed a memorandum of understanding. As a result of intervention by CIRM officers, the strike as well as the consequential estimated loss to the tune of Rs.100 crore could be averted.

3. Banking Sector

United forum of bank unions comprising of a Federation of trade unions/ associations of Bank employees/officers served a strike notice on chairman Indian Bank Association, Mumbai proposing to go on strike in the Banking Industry throughout India on 5th & 6th October, 2004 to press early settlement of long term wage revision due from 01.11.2002. The Chief Labour Commissioner (Central) held two rounds of conciliation proceedings on 22nd September & 1st October, 2004 and succeeded in persuading Indian Bank Association to increase the package of offer and the unions to withdraw the strike notice. As a result two day proposed strike in banks on 5th & 6th October, 2004 and consequential loss to the tune of Rs.322 crore approximately could be averted.

4.18 Industrial relations Machinery-Bi-partite agreement and Tri-partite agreements

Bi partite Agreement in signed under section 18 (1) of the Industrial Dispute Act. Which is applicable to those Settlement party only.

Tripartite Agreement in signed under section 12 (3) of Industrial Dispute act. This is applicable to presently employed, future employee, and all unions even if they are not a settlement party. Several Industrial Tripartite Committees (ITCs) have been constituted with the view to promote the spirit of tripartism. These tripartite bodies aim at deliberating on the industry-specific problems related to workers in the Cotton Textile Industry, Electricity Generation and Distribution, Jute Industry, Road Transport and Engineering Industry. Meetings of the Industrial Tripartite Committees on 'Road Transport', 'Jute Industry' and 'Cotton Textiles Industry' were held on 10th April, 25th August and 15th December, 2003, respectively.

The Government's proactive role through timely and effective conciliation of industrial disputes and involvement of social partners in the tripartite forums has successfully harmonized the interest of the employers and the workers resulting in a change of attitude from confrontation to that of co-operation.

4.19 JOINT CONSULTATIVE MACHINERY- An Introduction

The history of the civil service staff relations is characterised by the recognition of the fact that employees at work like to be treated as human beings. The recognition of the human dignity was one of the cardinal factors that lead to the resolutions of the grievances of the employees in a peaceful manner, in democratic method and at regular periods. The early employer-employee relations in the government service followed the traditional pattern, where the government employees were expected to owe unquestionable loyality to the State. Therefore, the conditions of service of the employees were determined unilaterally by the government without consulting the employees. The labour in the private sector secured benefits in matters of wages and working conditions as a result of effective labour organisation and militant actions. This led to the conviction among government employees' organisations that through joint and united efforts and concerted action only they could ensure improvement in their service conditions. Accordingly, the government employees resorted to strikes and brought pressure on the government to negotiate in good faith with its employees. They demanded the State' to be a model employer for promoting sound employer-employee relations.

Modern democratic governments with large-scale administrative organisations committed to the all-round development of the society cannot carry out their policies and programmes without the active cooperation and participation of the employees. Therefore, harmonious staff relations are absolutely necessary for ad3ministrative efficiency and it is not possible for the governments to maintain cordial relations without an effective form of consultation and negotiation. For this reason, the civil service staff relations policies and programmes must be consistently blased on the principles of democratic government Strikes by the government employees, it was realised, cannot be prohibited without providing suitable framework of joint consultation, collective negotiation procedures and methods of dispute settlement by compulsory arbitration.

As such, establishment of a grievance redressal and disputes solving machinery was considered the "essential half-way house" between the unilateral imposition of conditions of service in the public employment by the state as employer on the onehand, and deadlocks and strikes on the other. It was felt that all matters relating to pay, hours of work and conditions of service may be negotiated by the representatives of the employees and the employer. Thus, the discussions in joint consultative bodies generally relate to the exchange of information and consideration of the suggestions for improving safety, security, health and welfare and increase of productive efficiency. :The results of such mutual discussions usually take the shape of recommendatiorls framing the final decision to the government. The emphasis in joint coilsultation is mainly on informal method and cooperation based on common interests and good faith to improve the working conditions of employees and to increase the efficiency of administration. The simple aim of such relationship is to facilitate and promote peaceful resolution of conflicting interests. The outcome of such a process is influenced by the comparative abilities of the parties involved to reach agreement. It is the power relationship between the employer and the employees. In this unit we will examine significance of the Joint Consultation in civil service staff relation, evolution of Joint

Consultative Machinery, functioning of Joint Consultative Councils and Board of Arbitration and also discuss suggestion for the improvement.

4.20 Evolution of Joint Consultative MACHINERY

It was the introduction of 'Whitleyism' in 1919 in the British Civil Service Staff relations hailed by several countries as the novel method in the area of human relations. The Whitley system which involves formal as well as informal consultations and negotiations leading to agreement or arbitration has been an example followed in a large number of countries of the British Commonwealth with certain modifications. The staff relations in the British Civil Service are maintained through Whitley councils method. The Civil Service National Whitley Council is responsible for joint consultation on conditions of service affecting civil servants in general irrespective of the department to which they belong. The Departmental Whitley Councils are responsible for staff matters within their departments. Local Whitley Committees discuss matters affecting the local working conditions and day-to-day problems. The Government of India and the organisations of employees struggled nearly for three decades to have staff relations on the pattern of Whitley Councils in the U.K. The issue was first raised in 1928 anti again in 1942. The Royal Commission on Labour in 1931 recommended the establishment of a joint standing machinery for Railways. However, the Home Department did not accept the proposals, in spite of the favorable attitude of the Labour Department. The government referred the matter to the Central (First) Pay Commission in 1946 to enquire on "The machinery for negotiating and setting the questions relating to conditions of service which may arise out of differences between government and its employees". The Commission recommended (1947) the establishment of Joint Councils for employees. excluding those in Clas's **I**, on the lines of Whitely machinery. The Commission also recommended arbitration in regard to Class II and Class III **Services resort to** ad hoc tribunals on the lines of Civil Services Arbitration Board in United Kingdom. However, the Railways and the **P&T** were not included in this Schemes they were covered by the Industrial Disputes Act, 1947. Though the government decided to establish a Whitely scheme in 1952, it was only in 1954 that a.11 the Ministries of the Government of India were asked to set up Staff Committees Staff Councils. However, the scheme was not extended to the Railways, P&T and the Ministry of Defence. The employees of these departments were considered as industrial employees. In these departments, staff relations were evolved independent of others. Permanent Negotiating Machinery (PNM) was established in Railways in 1952. Monthly Meetings were introduced in P&T in 1949 and Joint Negotiating Machinery was set up in Defence in 1954. As per the directions of the government each Ministry was required to establish two separate Staff Committees one for staff other than Class IV and another for Class IV employees to secure the greatest measure of cooperation between the government and its employees. The Staff Committees were advisory bodies.

The Staff Committee in respect of other than Class IV employees consist of:

- A Chairman, who was Secretary or Joint Secretary of the Ministry, nominated by the Ministry.
- Government representative were nominated by the Ministry from its officers not below the rank of Under Secretary.
- The staff were represented by one member each for every 20 members employed in the Ministry in the grade of Section Officer, Assistant, Stenographer and Clerk, elected from amongst the members of that grade for a period of one year.
- The Chairman in consultation with the representatives of the staff, nominated one of them as Secretary of the Committee.

With regard to the Staff Committee of Class IV employees.

- A Deputy Secretary of the Ministry was nominated to act as the Chairman of the Committee.
- The Ministry nominates its representatives on the Committee not below the rank of an Assistant.
- The members of the staff were represented by one representative each for every 20 members from the groups consisting of Record Sorters, Jamadars, Peons, Farrashers, Chowkidars and Sweepers. They were elected for a period of one year from amongst the members of grades in each group employed in the Ministry.
- The Secretary was nominated by the Chairman from amongst the Staff representatives for a period of one year.

Both the Committees were to meet at least once in three months. The quorum for a meeting was one-third of the representatives of the members of the staff. Discussions were held in the meetings on agenda circulated in advance. The decisions were recorded and sent to the concerned Ministry for necessary action.

Objectives and Functions

The objectives of the Staff Committees were:

- To consider suggestions for providing the standards of work,
- To provide a machinery to the staff for making their points of view known to the Government on matters affecting their conditions of service, and
- To provide means on personal contacts between officers and staff in order to develop cordial relations between them and to encourage them to take keen interest in their work. The functions of the Committees were to discuss matters relating to the conditions under which the staff were required to work, general principles regulating conditions of service, the welfare of the staff and

improvement of efficiency and standards of work. Discussions on individual cases were not permitted. The working of the Staff Committees proved to be unsatisfactory. The problems faced by many Ministries were that in the majority of the cases, decisions could not be arrived within the Ministries and the recommendations of the Committees very often had to be referred to the Ministries of Home Affairs, Finance, Works, Housing and Supply for their clarification and clearance.

Functioning of Staff Councils

In 1957, the Cabinet appointed a Sub-committee to review the functioning of Staff Committees. On the recommendations of the Sub-committee, some instructions were given to all the Ministries to:

- Re-name the Staff Committees as Staff Councils.
- Dispose quickly the references emanating from the Staff Councils by the House Keeping Ministries.
- Refer unresolved matters to a Coordinating Committee consisting of representatives of Ministries of Home Affairs, Finance, Works, Housing and Supply.
- Explore the possibilities of delegating more powers to the administrative Ministries to avoid delay in making references to house-keeping agencies.
- Appoint a Chief Welfare Officer in the Home Ministry and a Welfare Officer in each Ministry.

As per the decisions of the Cabinet sub-committee the Staff Committees were re-named as Staff Councils without any change in their constitution, objects and the rules of procedure. A Coordinating Committee consisting of representatives of the Ministry of Home Affairs, Finance, Works, Housing and Supply and the concerned ministry was constituted to decide matters which remained unresolved in the Staff Councils. The Ministry administratively concerned with Staff Councils decide what matters should be brought before the Coordinating Committee.

The Staff Councils during their existence (1954-59) made many recommendations. Most of them were accepted and implemented by different Ministries. But they were unimportant items like water coolers, transport facilities, grant of leave, office uniforms, canteen facilities, dispensaries and first aid boxes etc. Neither the machinery of administration was improved nor the important grievances of the employees were redressed. The Second Pay Commission report observed "due to lack of will on the part of the Government to accept the principles of joint consultation failed as an effective instrument for prevention of the disputes."

Therefore, the Second Pay Commission proposed that "the situation required the 'establishment of machinery which fully in spirit and largely in form followed the

Whitley Machinery in the United Kingdom". Thus, the Staff Committees and Councils had "little in common with the Whitley Machinery" and failed to square up with the real Whitley spirit.

In India, the government civil employees have not shown any remarkable preference for joint consultative machinery. On the other, the government did not consult the organisations of the employees when it introduced the Staff Committees and Councils. It was a unilateral attempt for a bilateral purpose. The Whitley cause also could not be served in the absence of arbitration machinery for resolving the disputed matters. Further, due to their advisory nature the scope of the Staff Committees/councils and their activities were much restricted and the way in which they were handled by the officers further limited their utility. There was no Central Joint Staff Council like the National Whitley Council to consider issues of common and general application to all the employees of the Ministries. As the Civil Service Unions/Associations were not involved, the organised sections of employees declared no faith, rejected the Councils and considered them as "merely eye-wash". The officials at the top level were responsible for the ineffective functioning of Staff Councils for want of Whitley thinking. The staff representatives on the Committees/Councils lacked the necessary leadership to represent their views effectively. In actual practice, they merely acted as forums for the staff to ventilate their grievances and put forward them to the nominees of the government in the Staff Councils.

4.21 Emergence of Joint Consultative Machinery (J.C.M.) and Compulsory Arbitration Scheme

As already, stated, the Second Pay Commission recommended (1957), a Whitley type machinery for the whole body of Central Government employees, both-industrial and non-industrial to negotiate disputes. A special Committee of the Central Joint Council dealt with matters peculiar to the industrial employees. The Commission also recommended, compulsory arbitration, as a necessary compliment to the joint machinery for negotiation in matters of pay and allowances, weekly hours of work and leave.

In 1960, the Central Government employees went on strike for five days from July 11-16. One of their demands was the establishment of joint consultative machinery (J.C.M.). Immediately, the Government decided to set up the J.C.M. and proposed a scheme. Accordingly, three-tier machinery with joint consultation at the Central, the Departmental and regional/local levels as supplementary to existing arrangements was ' proposed. The scheme covered all the civil employees of the Central government including the P&T Department, Civil Aviation Department and the Ministry of Defence. The Scheme proposed an arbitration tribunal with a limited compulsory arbitration provision. The government during 1960-61 proposed to bring two bills before the Parliament to give statutory effect to the proposed J.C.M. and ban strikes by government employees. The employees' organisations protected against the proposals of the government and refused to give up right to strike and disassociate outsiders from their organisations. After discussions between the government and the employees organisations the Government of India decided in 1963 to introduce a scheme for joint consultative machinery and compulsory arbitration. On many aspects of the scheme the employee's organisations expressed doubts and opposed certain proposals of the government. The government wanted to bring all the employees (including industrial employees) under one umbrella. Therefore, there were discussions on many occasions at all levels for three years and all the doubts were clarified and there was a general consensus that a fair' trial be given to the scheme. Thus, the Joint Consultative Machinery and Compulsory Arbitration Scheme for Central Government Employees was inaugurated on October 28, 1966. The inauguration of the Scheme of J.C.M. was considered "opening of a new chapter" and hoped that the relationship of Governments with its employees would enter a new era of more fruitful cooperation.

4.22 The Salient Features of J.C.M. Scheme

The Scheme was designed "with the object of promoting harmonious relations and securing the greatest measure of cooperation between the government and its employees in matters of common concern, and with the further object of increasing the efficiency of the public service combined with the well-being of those employed" The Scheme is a voluntary one. The government and the employees unions and associations who participate in the scheme are required to subscribe to a Declaration of Joint Intent. Accordingly, both the sides (the government and the staff associations and unions) agreed to full and frank discussions on all matters in the Joint Councils to reach agreements.

The unique feature of the Scheme is its coverage of both industrial and non-industrial of the government. In this regard it differs from the Whitley System which is concerned with only non-industrial civil service. Another difference is that the J.C.M. Scheme does not cover "the class I, class II services, employees of industrial establishment and the union territories and police personnel" Yet another difference is that the fundamental objective of Whitley System viz., to provide machinery for dealing with grievances and to bring together experience and points of view of representatives of different classes and grades of civil service is not stated in the J.C.M. Scheme.

The J.C.M. provides for a three-tier structure and Joint Councils at the National, Departmental, Regional /office levels. The J.C.M. Scheme is a bi-partite body consisting of the representatives of the government (official side) and the representations oft he recognised employees organisations (staff side). The official side is nominated by the government and the staff side seats allocated to different **associations/unions/ federations** recognised by the government.

The Scheme broadly covers about 3 million regular class III and IV civil employees of the Central Government including industrial employees working in the departmentally run undertakings like the Railways and workshops and production units of various ministries. Though the Indian J.C.M. Scheme was modeled after the Whitley System in

U.K., the Indian Scheme could not inculcate in its scheme, the spirit and the long experience of the Whitley system. It is more a joint consultative machinery than a bi-partite participative agency.

4.23 The Functions of Councils

The Joint Councils deal with all matters concerning the conditions of work, standards of work, efficiency and staff welfare. However, in matters of recruitment, promotion and discipline, consultation is limited only to matters of general principles and individual cases are not considered. The Councils may appoint Committees to study and report on any matter falling within t5eir scope. All the agreements reached between the official and staff sides of a council will become operative subject to the final authority of the Cabinet. If the matter is arbitrable and final disagreements were to be recorded it may be referred to arbitration, if either **side** desires so. However, compulsory arbitration is not available at regional/ office council level.

National Council

The National Council is the apex body. It is one of the largest joint councils consisting of 85 members. The official side with a maximum membership of 25 is nominated by the government.

The staff side is nominated by the recognised **federations/unions/associations** of the employees. The seats are distributed between the **federations/unions/ associations** by the Chairman of the council. The distribution of seats is based on the numerical strength of staff employed in each Ministry/Department.

The Cabinet Secretary is the Chairman of the Council. He is the leader of the official side and is connecting link between the government and the employees.

The staff side elects one of its members as a leader by simple majority for a term of one year. The official and staff sides appoint Secretaries from amongst their representatives. There is a permanent secretariat of the council, which functions under the control of the Chairman.

The National Council deals with matters generally affecting Central Government Employees such as minimum remuneration, dearness allowance, and pay of certain common categories like the office clerks, Peons, and the lower grade of workshops and matters relating to categories of staff common to two or more departments but not grouped into one departmental council. The National Council would not deal with matters pertaining to a single department.

The National Council ordinarily may meet as often as necessary but not less than once in four months. A special meeting may be convened by the Chairman on his own or at the request of either official or staff side. The quorum for a meeting is one-third of the individual strength of the official and staff side.

The National Council may constitute two Standing Committees one for the industrial employees and the other for the non-industrial staff to deal with their respective matters. The Council may delegate any of its powers to the Standing Committees for the quick disposal of its business. The Council and the Committees may also appoint Sub-committees from amongst their members to study and report on any matters falling within its jurisdiction.

A matter once disposed of by the council cannot be brought to the agenda during the following 12 months, unless the Chairman permits it for any special reason. All matters should be decided by the council without reserving them for a later decision by the government.

The success or failure of the J.C.M. Schemes mainly depends upon the success or failure of the functioning of the National Council. For, the National council is expected to function as a machinery to redress grievances at the national level dealing with matters affecting all the employees of the Central Government.

Since its inception in 1966 it had held 25 meetings till the end of 1988. Over a period of 22 years of its existence it could withstand all the problems though it had some failures. In the initial years, the council could not reach certain agreements and the employees went on a day's token strike in September 1968 protesting against unhelpful attitude of the government. Due to the strike many of the employee's organisations were de-recognised for a couple of years. Therefore, the National Council could not meet with full strength. However, the council and its Committees were very active during 1974-76 to consider the recommendations of the Third Pay Commission.

Though the National Council could not succeed in achieving all its objectives, it acted effectively as machinery for dealing with the grievances of the employees and reduces the amount of conflict with the government. However, the National Council has come to stay and has taken roots to provide a system of joint consultation and cooperation.

Departmental Councils

Under the J.C.M. Scheme there is one Departmental Council for each Department, The scope and functions of these councils include all matters relating to the conditions of service and work, welfare of the employees, improvement of efficiency and standard of work. Councils deal with the problems of employees working in a Ministry/Department and the subordinate and attached offices of the department.

According to the J.C.M. Scheme 21 Departmental Councils are to be constituted in various departments. Till 1974, only 14 Departmental Councils were constituted. In the initial period it became difficult to constitute the Departmental Councils for want of recognised unions/associations.

The Secretary of the Ministry represents the official side and is the Chairman of the Departmental Council. The official side is nominated by the government or the Head of the Department. The strength of the official side is ten. The Chairman may invite temporary members and experts to the meetings for their advice.

The Associations/Unions recognised by the Department nominate their representatives for a term of three years on the staff side. The staff side elects one of its members as its leader for a period of one year. There is a permanent secretariat of the council under the control of the Chairman.

The membership of the staff side varies from twenty to thirty depending on the total strength of the employees and the number of grades and services in the department.

The Departmental Councils ordinarily meet as often as necessary and not less than once in three or four months. The quorum is one-third of each of the strength of the official and staff sides. The Councils may appoint Committees on ad hoc basis.

Regional/Office Councils

The J.C.M. Scheme permits the departments for the setting up of Regional/office Councils, where it is possible. The Councils deal with only regional or local matters. The strength of a regional/office council is determined by the size of the staff in a region or office. The Head of the region or office is the Chairman of the Council. The J.C.M. Scheme is silent about the constitution and the number of councils. Much progress is not recorded in the establishment of the councils, as there are some doubts in the minds of the Departmental authorities.

The matters to be discussed are limited to subjects within the competence of Joint Secretary (Administration) of a Ministry/Department such as revision of duty hours, distribution of work, accommodation, amenities, holiday duties, etc.

The Office Council consists of not more than five members on the official side and eight members on the staff side. The Joint Secretary (Administration) is the Chairman of the Council and the Under-Secretary (Welfare) is the Secretary of the official side.

The staff side is nominated by the recognised associations of the staff. The staff side elects its leader for a period of one year and appoints a Secretary from amongst its members.

The office council meets at least once in two months. The quorum is 113 of the members of each side. It may appoint committees to study and report.

If the council cannot arrive at an agreement on any issue it may be brought before the Departmental Council concerned. Arbitration is not available at office council level in case of disagreement between two sides.

To realise the full objectives of J.C.M. Scheme, setting up of regional/office councils is very important. For, they are the training fields to the representatives of the staff and official sides and provide grassroots to the J.C.M. Scheme. However, much progress is not achieved in establishing these councils for want of organisation of employees and positive attitude of officials at the local levels.

Summary

- Industriai relation is a dynamic concept which depends upon social changes, economical changes and changes in political setup of a country.
- Industrial relation is also a relative concept as it flourishes and damps relative to the changes in the economic, social and political conditions prevailing in a society and the laws made by the State to regulate them.
- Approaches to Industrial Relations are- Dunlop's approach, oxford approach, Industrial sociology approach, the action theory approach, the Marxist approach etc
- After independence in India government has been playing a comprehensive and vital in stopping out the structured pattern of industries
- Central industrial relations machinery in controlled by the chief labour commissioner
- CIRM perform actions like enforcement like labour laws Maintaining industrial relations and trade commission membership verifications
- Joint consultative Machinery was introduced by Whiteman in 1919 in British civil service staff relations
- Strategies of industrial relations are employee synergy consolidation of bargaining, internal leadership, empowerment , compensation Strategy etc.

Objective type Questions

1. Dunlop's approach was developed by-

   (a) John T Dunlop (b) Richard Arkwright

   (c) Karl Marx (d) Francis Bacon

2. Actors-Rules-Ideology are elements of -

   (a) Dunlop's Approach (b) Action theory Approach

   (c) Weber's Social action theory (d) Marxist Approach

3. The relation G.R=f (C.B) , where G.R= Governing rules and C.B=Collective bargaining was given by-

   (a) Dunlop's Approach (b) Action theory Approach

   (c) Weber's Social action theory (d) The Oxford Approach

4. Relationship between the management and workers of an industry is called as -

   (a) Human relation (b) Social relation

   (c) Industrial relation (d) Political relation

5. According to Industrial sociologist G. Margerism, the core of Industrial relations is the nature and development of the ____________ itself.

   (a) Peace (b) Conflict

   (c) Political harmony (d) Grievances

6. Which organisation is controlled by Chief Labour Commissioner?

   (a) CIRM

   (b) ISRO

   (c) International Labour organisation

   (d) Ministry of Labour and employment

7. The "institution of job regulation" is related to which approach-

   (a) Oxford Approach (b) Pluralist Approach

   (c) Marxist approach (d) Dunlop's Approach

8. Joint Consultative Machinery was introduced in which year?

   (a) 1900 (b) 1915

   (c) 1919 (d) 1920

Review Questions

1. Discuss the Concept and scope of Industrial Relations.
2. Write Short Notes On

   (A) Dunlop's Approach

   (B) The Oxford Approach

   (C) The Marxist's Approach

   (D) The Gandhian Approach
3. What are the Industrial relations Factors that influence the behavior of the people?
4. Mention the various objectives of Industrial Revolutions.
5. What are the aspects of Industrial relation? Mention the effects of poor industrial relations.
6. Discuss the difference between Industrial Relations and Human Relations.
7. What are the future strategies of Industrial Relations in India?
8. Discuss in Brief, the Role of Government in Industrial relation.

9. Discuss the evolution of Joint Consultative Machinery
10. What are the salient features of Joint Consultative machinery scheme?
11. Differentiate between Bi-Partite and Tri Partite agreements
12. Enlist various functions of CIRM.

Chapter 5

Grievance and Conflicts

5.1 An Introduction

A "grievance" is a written complaint. It can be about how the employer is using the collective agreement. An employee may also disagree with the employer about what the agreement means, and file a grievance about that. As well, a grievance can be used if the employee thinks that the terms and conditions of their job have changed unfairly, or the same not being honoured.

A "grievance" is not a tool for dealing with suspensions, dismissals, or competition appeals. Nor should this particular process be used to file harassment or discrimination grievances. All these issues are properly addressed through separate process with the assistance of a service officer or other designated union representative. Grievances are also not the right tool for day to day discussions and the occasional disagreements that can occur between employees and their supervisors.

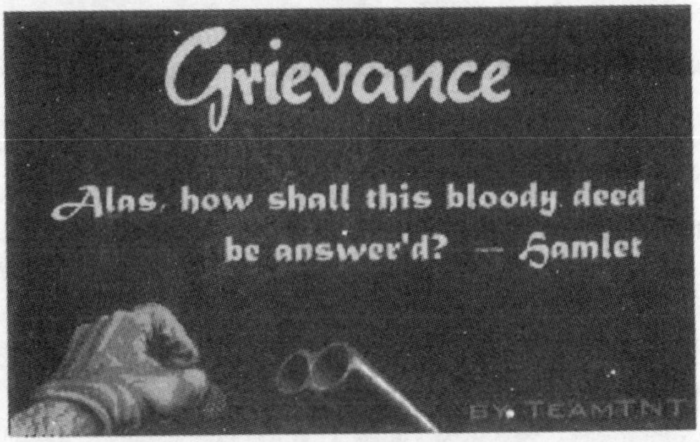

Some Quotes on Grievances by different sociologists are.

Grievance may be defined as "a written complaint field by an employee claiming unfair treatment".

Dale Yader

Grievances are complaints that have been formally registered in accordance with the grievance procedure.

Jackson

Complaints affecting one or more individual workers in respect of wage payment, overtime, leave, transfer, promotions, seniority, work assignment and discharge constitute grievance.

National Commission of Labour

A grievance is any dissatisfaction or feeling of injustice in connection with one's employment situation that is brought to the attention of the management.

Beach

5.2 Causes of Grievances

Following are some aspects which results in occurrence of grievance:

- Complaint against discipline, administration and behavior of supervisors.
- Violation of rules or agreements.
- Working condition of employee like safety and health.
- Alcoholism, absenteeism and accidents in any organisation.
- Complaint about job classification and incentive system.
- Violation of Laws of Central or State Government.
- Loss of seniority or calculation/interpretation of seniority.
- Unscheduled administrative changes.
- Harassment of union management relations.
- Promotion, demotion and transfer policies.

5.3 Effects of Grievance

Some of the important effects of Grievance's are mentioned below:

- Decline in production and productivity.
- Increment is production per unit cost.
- Strikes and lock outs.
- Increase in incidence of accidents and safety hazards.
- Reduction is level of discipline like commitment, sincerity and punctuality.
- Increase in incidence of accidents and safety hazards.
- Wastage of available resources.
- Increase in unrest.
- Conflicts and disagreements.

- Poor efficiency of workers due to lack of interest in their respective field.
- Increases in level of supervision and control.
- Strain between superiors and their subordinate.

5.4 Types of Grievances

A grievance may be specifically characterized into 4 major types:

I. Disagreement over Meaning or Application

In this type of grievance the facts of the issue are not usually in dispute. Something did or did not occur. A contrary meaning is attached to a term or condition of the agreement by the union and the administration. Ex.□ days allotted for death in the family, the administration does not consider a first cousin (blood relation) family.

In these instances, the minutes from negotiations must be taken into account to determine the understanding of the negotiation members.

II. Plain Violation of the Agreement

This type of grievance may be the result of ignorance, carelessness, error, omission of an act known to be contrary to the terms of the agreement. This type is the simplest type of grievance to substantiate.

Ex. The administration schedules a work day that is from 7:30□3:30pm. (8 hours long) This matter can be substantiated by gathering evidence to support the length of the day that the employee worked. The administration is not contesting the interpretation of this term of the agreement but is rather ignoring it.

III. Dispute over the Facts

In this type of grievance there is no question as to what the agreement provides. The issue turns on whether the act which prompted administrative action occurred or did not occur.

Ex. All employees must be in their rooms 20 minutes before student start time. A written warning is given to a teacher who claims she was in the room at the designated time. The case rests on whether or not the employee was in the room. The solution is easier if the employee can document her information. However, the administrator took the initiating action and he/she would have the burden of supporting the contention.

IV. Disagreement as the Reasonableness of Administrative Acts

In this type of grievance, the union feels the administrator exceeded his discretionary authority under the agreement.

Ex. The contract states: "Extra duty is to be assigned on a rotating basis."

The employee feels he has been assigned duty more frequently than any other teacher.

The central issue here is the seeking of equity□ the application of the terms and conditions of the contract in a fair and impartial manner.

5.5 Grievance Handling Procedure

It is a method formulated by managerial authorities to deal with redressal of grievance. This handling procedure may vary from organisations to organisation. Pettefer in 1970 payed the four steps for handling Grievance.

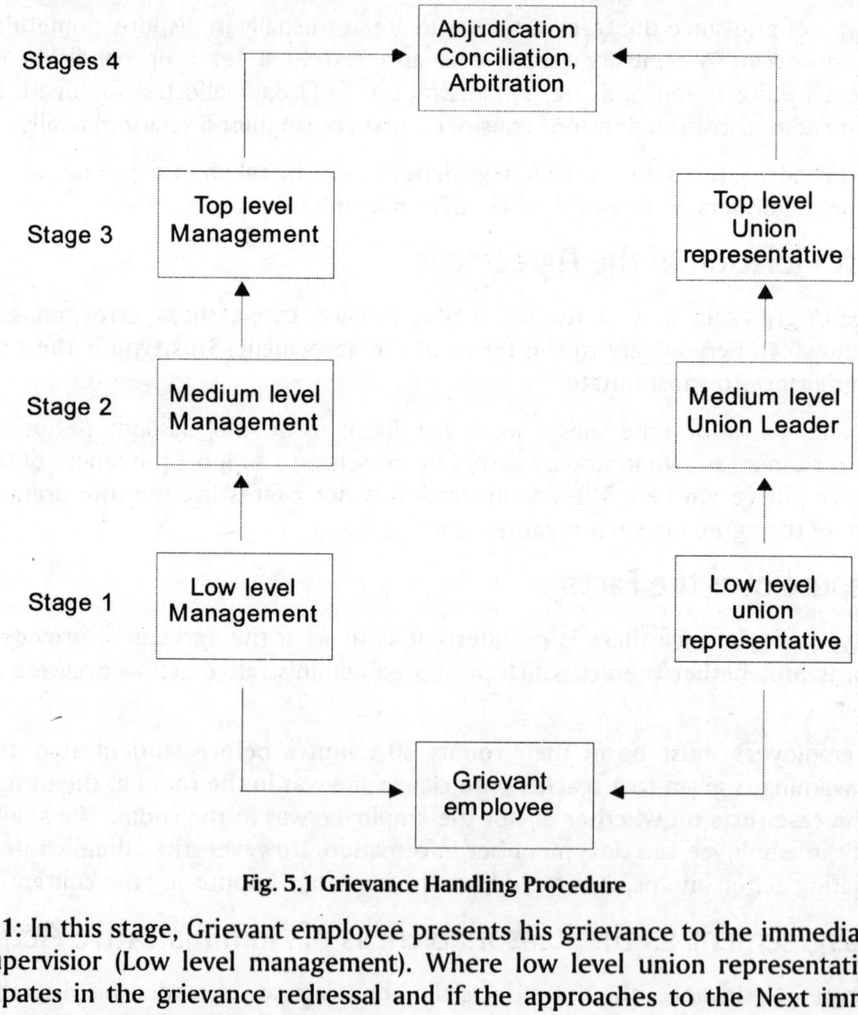

Fig. 5.1 Grievance Handling Procedure

Stage 1: In this stage, Grievant employee presents his grievance to the immediate boss i.e. supervisior (Low level management). Where low level union representative also participates in the grievance redressal and if the approaches to the Next immediate senior management level for settlement.

Stage 2: The grievance which was not solved by the former management, is settled in this level usually middle order managers and middle order union leaders participates in settlement of Grievance. If again the settlement or we can say satisfaction is not achieved by the grievant them, it may approach to the next level, which in second last level.

Stage 3: In this stage, top level management officials and top level union leader may be president or secretary participates in settlement of grievance. Members involved in this stage are highly qualified and experienced in solving grievance matter. It again, the settlement is unsatisfactory then grievant approaches to next level which is final stage.

Stage 4: In this stage, third party named as arbitrator plays a vital role, who is supposed to give an impartial decision Arbitrator settles the grievance matter and if possible then refers to the Labour court for settlement. This stage involves conciliation, arbitration and adjudication. In case of labour court decision processed will be final and applicable to both the parties and are subjected to statutory provisions for high court's appeal.

5.6 Grievance Handling: Precautions and Prescriptions-

Introduction

Maintaining quality of work life for its employees is an important concern for the any organisation. The grievance handling procedure of the organisation can affect the harmonious environment of the organisation. The grievances of the employees are related to the contract, work rule or regulation, policy or procedure, health and safety regulation, past practice, changing the cultural norms unilaterally, individual victimization, wage, bonus, etc. Here, the attitude on the part of management in their effort to understand the problems of employees and resolve the issues amicably have better probability to maintain a culture of high performance. Managers must be educated about the importance of the grievance process and their role in maintaining favorable relations with the union. Effective grievance handling is an essential part of cultivating good employee relations and running a fair, successful, and productive workplace. Positive labor relations are two-way street both sides must give a little and try to work together. Relationship building is key to successful labor relations.

Precautions and Prescriptions

The management should take care of following aspects to develop a culture of trust and confidence upon the employees.

- Always ensure that the managers involved in the grievance handling procedures have a quiet place to meet with the complainant.
- Always ensure that managers have adequate time to devoted to the complainant.
- Explain manager's role, the policy and the procedures clearly in the grievance handling procedure.
- Fully explaining the situation to the employee to eliminate any misunderstanding and promote better acceptance of the situation complained of.

- Try to let employee present their issues without prejudging or commenting
- Do use a positive, friendly ways to resolve the crisis than punitive steps, which disturb the system.
- Do remain calm, cool, collected during the course of the meeting.
- Always focus on the subject of the grievance than allied issues.
- Don't make threats manage the grievances.
- Never make use of allegations against personalities.
- Be aware of the staff member's potential concerns to the possible repercussions of raising a grievance.
- Don't become angry, belligerent, or hostile during grievance handling procedure.
- Do listen for the main point of arguments and any possible avenue to resolve the grievance.
- Listen and respond sensitively to any distress exhibited by the employees.
- Eliminating the source of the irritation or discomfort being complained of.
- Reassure them that the managers will be acting impartially and that your hope is to resolve the matter if possible.
- Don't "horse trade" or swap one grievance for another (where the union wins one, management wins one). Each case should be decided on its merits.
- Avoid usage of verbalisms like "it will be taken care of."
- Ensure effective, sensitive and confidential communication between all involved.
- Take all possible steps to ensure that no victimization occurs as a result of the grievance being raised.
- The investigator or decision maker acts impartially, which means they must exclude themselves if there is any bias or conflict of interest.
- All parties are heard and those who have had complaints made against others are given an opportunity to respond.
- Try to look upon the problem from different angles for appropriate understanding.
- Ensuring that there is proper investigation of the facts and figures related the problem under concern.
- Consider all relevant information in the investigation process.
- Ask the staff member their preferred resolution option, although it is important to make it clear that this may not be a possible outcome.

- Be aware of the limits of authority of the person who involved in the grievance handling procedures.
- If the manager feels that he/she is not the appropriate person (senior manager) to deal with the issue refer the complainant to the appropriate person as soon as possible.
- Try to get a better idea of whether the alleged discrimination or harassment happened or didn't happen.
- Tell them exactly what they are supposed to have done, to whom and explain, why this may be seen as discrimination/harassment or as inappropriate.
- Grievances are preferably to be settled informally at the level of the employee's immediate supervisor.
- Try level best to involve team members to resolve the crisis at unit level itself.
- Avoid as far as possible the union involvement in conflict resolution situation process.
- Follow the documentation procedures, of all necessary steps taken to resolve the problem/complaint.

Conclusion

To a great extend the aggravation of industrial problems depends on manager's approaches and attitude in effective handling of employees grievances. Care should be taken in the way managers approaches the problem and perceiving the pros and cons of the situation. The conflict management approaches include the win-win strategy that help in the healthy organisational practices and which reflects the strong organisational culture. The cooperation from both parties is the pre-requisite to handle the problem and effective settlement of the grievances. Conscious use of professional self can help managers in the conflict handling situations grievance redressal process.

5.7 Individual Employee Grievance Procedure

This procedure shall apply to all members of staff other than "holders of senior posts" as defined in the College's Articles of Government. The procedure aims to help to resolve individual grievances in a manner which is as fair and expeditious as possible. It is the College's policy to find a solution to individual grievances as early in the procedure as possible.

Where separate procedures exist which deal with complaints relating to e.g. harassment, bullying, health and safety issues etc., the complaints should be dealt with in accordance with that procedure. The outcome of that procedure may result in referral to the grievance and/or disciplinary procedure at an appropriate stage.

A separate procedure also exists for the handling of Collective Disputes.

Employees who have a grievance or those against whom a grievance is raised have the right to be accompanied and represented by a trade union representative or work place colleague of their choice at every stage. Every effort will be made to resolve the grievance at the informal stage.

In certain circumstances and where mutually agreed following consultation with Personnel, it may be necessary for earlier stages to be bypassed and the procedure to commence at a later stage.

If the grievance is against the immediate line manager, then the person to whom the grievance should be sent should be senior to the line manager.

If the grievance is against the Principal, it should be sent to the Clerk to the Corporation (email address: brownr@phoenix.deeside.ac.uk)

Stage 1 - Informal Discussions

1.1 If an employee has a grievance relating to his/her employment, the matter should be raised initially with the employee's immediate supervisor/Head of department as appropriate. The grievance should be raised orally in the first instance. In the event that the grievance relates to the employee's immediate supervisor/Head of Department who would normally deal with a grievance at this stage, the grievance should be referred to the employee's immediate supervisor's own manager/Head of Department's immediate manager.

1.2 The appropriate supervisor or Head of Department will attempt to resolve the complaint informally. He/she shall enquire into the grievance and will discuss it with the complainant and will normally communicate the decision to the complainant orally within 5 working days after the complaint is received. In the event that the complainant does not find the decision acceptable, he/she may request the decision in writing, which will be provided within 2 days. Most grievances will be resolved at this stage.

Stage 2

2.1 If the grievance has not been resolved at Stage 1 the complainant may refer the grievance to the Head of Department/Vice Principal as appropriate by obtaining from the Personnel Department, and completing, a grievance form. The complainant must complete the form by providing full details of the complaint and should address it to the Human Resource Manager.

2.2 The Human Resource Manager will, as soon as possible [and in any event, within five working days of receipt of the form] arrange a meeting at which all parties to the grievance will attend. The employee may be represented and accompanied at the meeting by a trade union representative or work place colleague.

2.3 The employee bringing the grievance or his/her representative will be given the opportunity to explain the nature of the grievance, submit verbal/written evidence and call appropriate witnesses.

2.4 The Head of Department/Vice Principal as appropriate will have the right to ask questions of the parties in attendance.

2.5 The Head of Department/Vice Principal as appropriate will consider the entire matters rose at the meeting and will issue a written decision within five working days. Copies of the decision will be sent to all of the parties.

Stage 3

3.1 If the complaint is not resolved to the satisfaction of the employee at Stage 2, an appeal may be submitted, in writing, to the Principal within ten working days of receipt of the decision reached at Stage 2.

3.2 The employee will be entitled to have a meeting with the Principal and to be accompanied and represented by a trade union representative or work place colleague.

3.3 The Principal will consider the appeal and may be supplied with all of the documentation submitted in relation to the earlier stages of the procedure.

3.4 The Principal will issue and send to all parties, copies of a written decision within ten working days of receiving the appeal in writing. Such decision will be final.

5.8 Understanding Grievance Procedure in Human Resource Management

If an organization has to move towards excellence, maintenance of harmonious and cordial relationship is a vital condition. Similar to organization' expectations from the employees. The employees do have more expectation in terms what they have contributed to each other. Failure to meet with each others expectation or the deviations from what has already been accepted may lead to indiscipline, grievance and stress are of continuing in nature and often judicial- legal process may not be of much help in resolving them.

Managing Discipline:

In the modern management process, discipline should be viewed as a behavioral modification process. Although there may be many norms fixed for desirable behavior wherever there are deviations. Appropriate corrective measures should also be taken for overall effectiveness, of an organization. Promotion, maintenance and strengthening of discipline will be a continuing affair, only if some of the following conditions are fulfilled. They are normally

- A shared culture
- Inculcation of positive attitudes
- Scope for problem solving and
- A feeling of satisfaction

(1) Approaches to Deal with Indiscipline:

This approach tends to emphasize the coercive and punitive methods that fall within the legal-constitutional framework. It may not always be realistic.

(2) Judicial Approach:

This is only an after effect approach. It follows the law of natural justice and provides the offender all possible opportunity to bring out his side of the case. This is a time consuming process and conducive climates are not often restored.

(3) Humanistic Approach:

It lays emphasis on a healthy inter personal relationship between the employer and employee. Corrective steps are taken in helping the employees to get over their difficulties and to the extent possible, punitive actions are avoided inspite of some perceived inconsistency that may exist among other workmen.

(4) Human Resources Development (HRP) Approach:

Keeping in mind the Theory Y and its implications, organizational goals have to be enhanced through adequate training, motivational patterns and personnel policies. Groups are used as influences and often made to act as catalysts to emulate norms of behavior. Likewise, the effectiveness in maintaining discipline is also possible through the process of leadership. A leader has to exhibit mutuality of interaction, persuasion, highly interpersonal in his relationships and get him involved and should expect participation from others to achieve the organizational goals and bear in mind not to indulge in favoritism while taking appropriate actions.

To avoid displeasure among the employees while enforcing discipline, it would be a desirable course of action if only one employees his diagnostic skills and intervention skills. If a change is to be expected in an individual's behavior, some amount of influence had to be exercised in order to create a congenial atmosphere.

Managing Grievance:

In a democratic set up any employee should be in a position to express his dissatisfaction, whether it be a minor irritation, a serious problem, or a difference of opinion in the work assignment or in the terms and conditions of employment. The feeling of discontent or dissatisfaction becomes a complaint when (a) it has not assumed (b) the matter has been presented in a highly informal way. A complaint turns into a grievance when (I) there is a feeling of injustice (II) expressing the feeling formally, either verbally or in writing and (III) it is related to policies, procedures and operations of the organization. The National Commission on labour (1969) for example defines grievance in the following way:

"Complaints affecting one or more individual .workers in respect of their wage payments, overtime, leave, transfer, promotion, seniority, work assignment and discharge would constitute grievance. Where the points at dispute are general applicability or of considerable magnitude they will fall outside the scope of this procedure".

Nature and Causes of Grievance:

In an organization a grievance may be presented by an employee or group employees, with respect to any measure or a situation that directly affects the individual or is likely to affect, the conditions of employment of many workers. If such a grievance is transformed into a general claim, either by the union, or by a group of employees, then the claim falls outside the scope of grievance procedure as it is a collective grievance and therefore it falls under collective bargaining.

In 'Grievance Procedure, a survey of practices in industries in India', it's classified into 19 causes of employee grievances have been outlined. They are

- Promotions
- Amenities
- Continuity of service
- Compensation
- Disciplinary action
- Fines
- Increments
- Leave
- Medical Benefits
- Nature of job
- Payment
- Acting promotion
- Recovery of dues
- Safety appliances
- Superannuation
- Suppression
- Transfer
- Victimisation
- Condition of work.

The international labour Organization (ILO), clarifies a grievance as a compliant of one or more workers with respect to wages and allowances, conditions of work and interpretation of service stipulation, covering such areas as overtime, leave, transfer, promotions, seniority, job assignment and termination of service

5.9 Grievance Management in Indian Industry

At present, there are three legislations dealing with grievances of employees working in industries. The Industrial Employment (Standing Orders) Act, 1946, requires that every establishment employing 100 or more workers should frame standing orders. These should contain, among other things, a provision for redressal of grievances of workers against unfair treatment and wrongful actions by the employer or his agents. The Factories Act, 1948, provides for the appointment of a Welfare officer in every factory ordinarily employing 500 or more workers. These Welfare officers also look after complaints and grievances of workers. They will look after proper implementation of the existing labor legislation. Besides, individual disputes relating to discharge, dismissal or retrenchment can be taken up for relief under the Industrial Disputes Act, 1947, amended in 1965.

However, the existing labor legislation is not being implemented properly by employers. There is lack of fairness on their part. Welfare officers have also not been keen on protecting the interests of workers in the organized sector. In certain cases, they are playing a dual role. It is unfortunate that the public sector, which should set up an example for the private sector, has not been implementing labor laws properly.

Guidelines for Handling Grievances

The following guidelines may help a supervisor while dealing with grievances. He need not follow all these steps in every case. It is sufficient to keep these views in mind while handling grievances.

- Treat each case as important and get the grievance in writing.
- Talk to the employee directly. Encourage him to speak the truth. Give him a patient hearing.
- Discuss in a private place. Ensure confidentiality, if necessary.
- Handle each case within a time frame.
- Examine company provisions in each case. Identify violations, if any. Do not hold back the remedy if the company is wrong. Inform your superior about all grievances.
- Get all relevant facts about the grievance. Examine the personal record of the aggrieved worker. See whether any witnesses are available. Visit the work area. The idea is to find where things have gone wrong and who is at fault.
- Gather information from the union representative, what he has to say, what he wants, etc. Give short replies, uncovering the truth as well as provisions. Treat him properly.
- Control your emotions, your remarks and behavior.
- Maintain proper records and follow up the action taken in each case.

5.10 Model Grievance Policy

A grievance policy and procedure should set out the steps to be followed to initiate and resolve grievances in the workplace.

Grievance procedures should cover issues such as health and safety, terms and conditions of employment, organisational change and equal opportunities.

1. Raising a Grievance Informally

Any employee who has a grievance relating to their employment may raise the matter informally with their manager during the course of normal work. This is not part of the grievance procedure. Only if the matter is not resolved between them should the formal grievance procedure need to be used.

2. The Formal Grievance Procedure

Stage 1

The employee should put their grievance in writing. The employee does not need to state that their complaint is a formal grievance; however it would be helpful if they would do so.

Stage 2

Once a written grievance has been received by the employer, then the organisation has a responsibility to investigate the matter and to invite the employee to a formal grievance hearing.

The employee may at this stage, if they wish, be accompanied by another employee/work colleague from the organisation or a trade union official or representative in accordance with the provisions of the Employment Relations Act 1999 when requesting that an issue be raised formally.

The supervisor is required to:

1. Record full details of the matter, or accept signed written details of the matter
2. Provide an answer in writing to the employee with a copy to the representative, if any. Should the employee then wish to take the issue to the next stage of this procedure, the supervisor shall notify the Stage 3 management representative without delay.

Under the statutory standard grievance procedure, a grievance hearing must not take place unless the employee informs the employer of the basis of their grievance. Also, such a hearing must not take place unless the employer has had a reasonable opportunity to consider their response to that information.

The timing and location of the grievance meeting must be reasonable. Therefore should an employee make a complaint with regards to an unacceptable working environment, the meeting would usually take place away from this environment and in most circumstances the meeting will take place during the employee's normal working hours.

Stage 3

The Stage 3 management representative will arrange to meet the employee concerned and/or, at the latter's wish, their representative.

At this meeting the employee may be accompanied as at Stage 1.

Following the meeting, the management representative is required to dispatch an answer, in writing, to the employee and, if applicable, to the representative.

The decision of the Stage 3 management representative will be final.

3. Time Limits

Under the Employment Relations Act 1999, if the employee's chosen companion (fellow worker or union official or representative) is not available on the day of the meeting the employee may reasonably request that the meeting be held on another day within five working days of the selected date.

The various steps of the procedure should be held within a reasonable time. Often the procedure will give an indication of these times but they should not be fixed, as it may not be possible to meet them. For example, if a detailed investigation is required, time will have to be extended.

In the absence, for any reason, of the appropriate management representative, the procedure may authorize any other executive of the [Company/Organisation] to take the necessary action on their behalf. At the request of the employee, however, consideration may be deferred until the return of the appropriate management representative.

Although there are time limits within which answers to grievances are to be given, it is important that grievances are dealt with as quickly as possible, and as close to the level of origin as possible.

5.11 Grievance Redressal Policy

1. Introduction

In the present scenario of competitive banking, excellence in customer service is the most important tool for sustained business growth. Customer complaints are part of the business life of any corporate entity. This is more so for banks because banks are service organizations. As a service organization, customer service and customer satisfaction are our prime focus. We believe that providing prompt and efficient service is essential not only to attract new customers, but also to retain existing ones. Our Bank has come up with a lot of initiatives that are oriented to providing a better customer service and a better complaints redressal mechanism with a view to **"Out serve"** customers.

The bank's policy, on grievance redressal has been formulated taking into account the following:

- Customers are treated fairly at all times
- Complaints raised by customers are dealt with courtesy and on time
- Customers are fully informed of avenues to escalate their complaints/grievances within the organization and their rights to alternative remedy, if they are not fully satisfied with the response of the bank to their complaints.
- All complaints are dealt efficiently and fairly
- The bank employees work in good faith and without prejudice to the interests of the customer.

In order to make bank's redressal mechanism more meaningful and effective, a structured system has been built. This system would ensure that the redressal sought is just and fair and is within the given frame-work of rules and regulation. All the employees are aware of the Complaint handling process. The customer has every right to register his/her complaint if he/she is not satisfied with the services provided by the **Bank or its service providers**. Customers can give their complaint in writing or over telephone or SMS or through the Internet Banking facility. In case the complaint is not resolved within the given time or if he/she is not satisfied with the solution provided by the bank, the customer can approach Banking Ombudsman or other available legal avenues with their complaint for grievance redressal.

2. Internal Machinery to handle Customer complaints/ grievances

Customers who wish to provide feedback or send in their complaint can use the following channels available with the Bank:

- Call our Phone Banking Help lines
- Email us at customer.care@SC.com or Write to us at Customer Care Unit, 19, Rajaji Salai, Chennai 600001.
- Avail 24 hours Service Assurance facility by sending SMS to our dedicated service number: 9980033333.
- "May I Help You" desks are available at all our branches. Customers can also approach our Personal Financial Consultants at the designated desks.

In case the response received through the above channels are not satisfactory, they can write to the Central Nodal Officer, Mr. Balu Jayaraman – Head, Customer Care at the address mentioned above or email him at head.service@SC.com. We assure a response to letters / emails received through this channel within 5 working days.

The information on the help line numbers and the redressal mechanism mentioned above are available in all our branches. Further, this has also been put up in our website and can be accessed through the following link:

http://www.standardchartered.co.in/personal/tools-utilities/en/complaints.html

The Bank also has dedicated monthly forums to discuss and analyze the causes for the complaint numbers, the complaint resolution process and to discuss the progress and action plans on the same. These detailed discussions of progress and action plans involve different levels of hierarchy that ensure a wide audience and the percolation of information to all the frontline units.

2.1 Customer Service Committee of the Board:

This committee comprises of senior officers of the Bank for examining any issues of the customers. The Committee examines any other issues having a bearing on the quality of customer service rendered. This Committee also reviews the functioning of Standing Committee on Customer Service. The committee is responsible for the following functions.

- To formulate the deposit policy
- To formulate product approval process
- Annual customer satisfaction survey
- To look at and decide on policy matters pertaining to Customer Service.
- To issue guidelines to the Standing Committee on matters relating to customers.
- To seek and obtain feedback from the standing committee on areas pertaining to customer service in terms of trends of complaints, service issues etc.
- Review and suggest way forward on the reports put up by the standing committee pertaining to complaints cases of delays in dealing with deceased accounts etc
- To have oversight over the implementation of RBI's circulars on customer service issued
- To look at all awards given by the Banking Ombudsman to address issues of system deficiencies brought out by the awards. Also look at reasons for awards which have remained unimplemented for more than 3 months if any.

2.2 Standing Committee on Customer Service:

The Standing Committee on Customer Service is chaired by Head of Service of the Bank.

Besides two to three senior executives of our bank, the committee also invites two to three eminent non-executives drawn from the public as members. The committee is responsible for the following functions:

- Evaluate feed-back on quality of customer service received from various quarters.

  The committee would also review comments/feed-back/ complaints on implementation of commitments in the Code of Bank's Commitments to Customers.
- The Committee is responsible to ensure that all regulatory instructions regarding customer service are followed by the bank. Towards this, the committee would obtain necessary feed-back from regional managers/ functional heads.
- The committee also considers unresolved complaints/grievances referred to it by functional heads responsible for redressal and offer their advice.
- The committee submits report on its performance to the customer service committee of the board at quarterly intervals.

2.3 Nodal Officer to handle complaints and grievances

Bank has appointed a Central Nodal Officer, Mr. Balu Jayaraman and he is responsible for the implementation of customer service and complaint handling for the entire bank.

3. Mandatory display requirements

Our bank has the following in all our branches;

- Appropriate arrangement of receiving complaints and suggestions.
- Display of the name, address and contact number of Central Nodal Officer(s)
- Contact details of Banking Ombudsman of the area
- Code of bank's commitment to customers

4. Resolution of Grievances

The customers can highlight their complaints / issues with our Bank vide the channels mentioned earlier in the policy. The officer in the concerned unit with whom the customer has raised the issue is responsible for the resolution of complaints/grievances.

The Branch Managers can also be contacted by the customers for lodging their complaints.

The officers of the complaints redressal unit will ensure closure of all complaints to the customers' satisfaction. They will ensure that the complaint is escalated to the appropriate levels in case it is not possible to resolve at his/her level. Whilst the ultimate endeavor is to ensure we reach a situation where our customers don't have to complain to senior management to get an effective redressal, we have put in a robust mechanism to handle these complaints, review them from a point of view of understanding reasons for the complaint and for the escalation and working on prevention of recurrence thereof.

4.1 Time frame

To register complaints, customers can use any of the channels mentioned above (refer point 2 on Internal Machinery to handle customer complaints). If the complaint has been received in writing, the bank will endeavor to send an acknowledgement / response within a week. After the matter is examined a final response will be sent to the customer or information that more time is required and the bank will endeavor to do so within 30 days of receipt of complaint. In case the customer is not satisfied with the response received, then he /she can write to: Mr. Balu Jayaraman, Head – Customer Care, Standard Chartered Bank, 19 Rajaji Salai, Chennai – 600 001 Ph: +91 044 – 64578822 (Monday to Saturday – 9AM to 6PM) Email: head.service@SC.com. The complaint will be responded within 5 working days. In case the customer is still not satisfied with the response or has not received a response from the bank within a month, then the customer can file a complaint before the Banking Ombudsman. More details on the banking Ombudsman Scheme available at www.bankingombudsman.rbi.org.in

Complaints that are received at our end will be seen in the right perspective and would be analyzed from all possible angles.

The communication of bank's stand on any issue will be provided to the customers. Complaints that require some time for examination of issues involved will be acknowledged promptly.

5. Interaction with customers

The bank, through various questionnaires / meetings / surveys obtains customer feedback / suggestions for improvement in customer service.

6. Sensitizing operating staff on handling complaints

All the staffs of the Bank are educated on our Complaint Redressal Mechanism. We are confident that with an open mind and a smile on the face we should be able to win the customer's confidence

Sanjog Helpline
e- Grievance Redressal Policy

Background:
Almost all state government schemes are citizen driven Citizen's interaction with Government is often overlooked. Many e-govt. projects in the way of grievance mechanisms simply act as cover-up technologies offering little advantage to citizens. A genuine complainant has no knowledge as to whom and how to address. Such situations give rise to touts, who register villager's problems for money. This apprehension of wasting time & money prevents a villager to vent his grievances. To them, the overall administrative system is autocratic and opaque.

Motivation:
The Directorate of Public Grievances (DPG) has a mandate to provide system where citizen is center and is consulted at various stages of public policy implementation and provide criticism. Thus empowered and enlightened, today's citizen is more demanding and the government, therefore, has to develop and enable itself to meet the evolving demands. DPG also promotes ICT as the medium of grievance registration and an aisle for performance assessment of government officials. Through ICT, various Govt. programs and their status can be assessed by the people and grievances can be looped in appropriate redressal forums.

As a precursor to its leadership efforts in e-Governance implementation, the Govt. of Orissa commissioned the Integrated, Multilingual Automated and web-based Grievance Redressal System to enable seamless registration of feedback of the citizenry, on-going implementation status and post-implementation impact of the welfare schemes instituted by the govt.

Process
The Panchayati Raj department commissioned Directorate of IT (OCAC), Orissa to develop a value based grievance redressal interface called Sanjog Helpline. The broad objective of the project was to enhance the process of grievance registration and redressal by providing a single platform to keep track of each grievance coming from the citizens. The overall objective was to provide an interface to the citizens through which they can register their grievances and at a later stage can track and acknowledge its status. On the other hand the system takes care of notifying the concerned Action Taking Authority (ATA) about the grievance registered for the particular scheme and location and keeps on following him up until and unless the grievance is solved. Citizens can register grievances through toll free number, fax, e-mail and through Sanjog Helpline portal from their villages.

The ICT tool of Sanjog Helpline ensures that the grievance of complainant is sent to the right office to the right officer in real time. It is developed on a web enabled multi-lingual platform which will receive complaints through its 9AM to 5PM Call Centre. This call center is responsible for entering the details of the complainant and the grievance in the web portal of Sanjog helpline at www.sanjoghelpline.in. This is the first foray into the online portal. Once a grievance is registered, the built in intelligent system ensures that it is forwarded to the appropriate action taking authority (ATA) for its redressal. The system automatically tracks and pursues the status of grievance with this ATA. The complainant through a ticket number can find out the status of his grievance through the portal anytime. Thus the web portal contains and maintains a database of all grievances registered, specific to location, nature of complaint and concerned authority of redressal.

Apart from this, the system can automatically escalate the complaint, to the seniors in the hierarchy of the particular scheme, if it is unresolved in the specified time period. This is made possible through a unique configuration module, where the department's and the scheme's hierarchy of officers is recorded into the system. In this process, it is ensured that every escalation that occurs, not only sends intimation over mail and SMS, but also sends a facsimile in a formatted government letter-head with the ink-signed signature of the senior. This process, not only ensures the authenticity of the job at hand, but also the implied seriousness of resolving the problem. It also benefits in the assessment of the performance of an ATA.

Once a grievance is closed or sorted, the web portal automatically updates its status. Therefore right from registering a grievance, a citizen, an ATA and any other officer in the hierarchy, can have the complete picture of the complaint and the entire mode of operations involved, along with the delays incurred while resolving the grievance.

Apart from this, Sanjog Helpline has a built-in MIS for administrators to identify the performance of their offices and officers therein for taking preventive actions aganist non-timely performance. Provision is also there to display reports based on fields, schemes, location, nature of complaints, modes and product so that it is helpful for government officials to take strategic decisions for the beneficiaries.

At present Sanjog Helpline caters to 5 schemes, namely NREGS, Rural Water Sanitation Scheme (specific to Tube well), PMGSY, Orissa Flood 2008 (Panchayati Raj) and SAMS (Students Academic Management System). Apart from these, there are initiations by other departments viz., Women & Child Development (WCD) Department and Health Department. The Orissa government has also issued a notice state wide for Sanjog helpline to be used as the State Grievance Redressal forum.

Workflow of Sanjog Helpline: How it works?

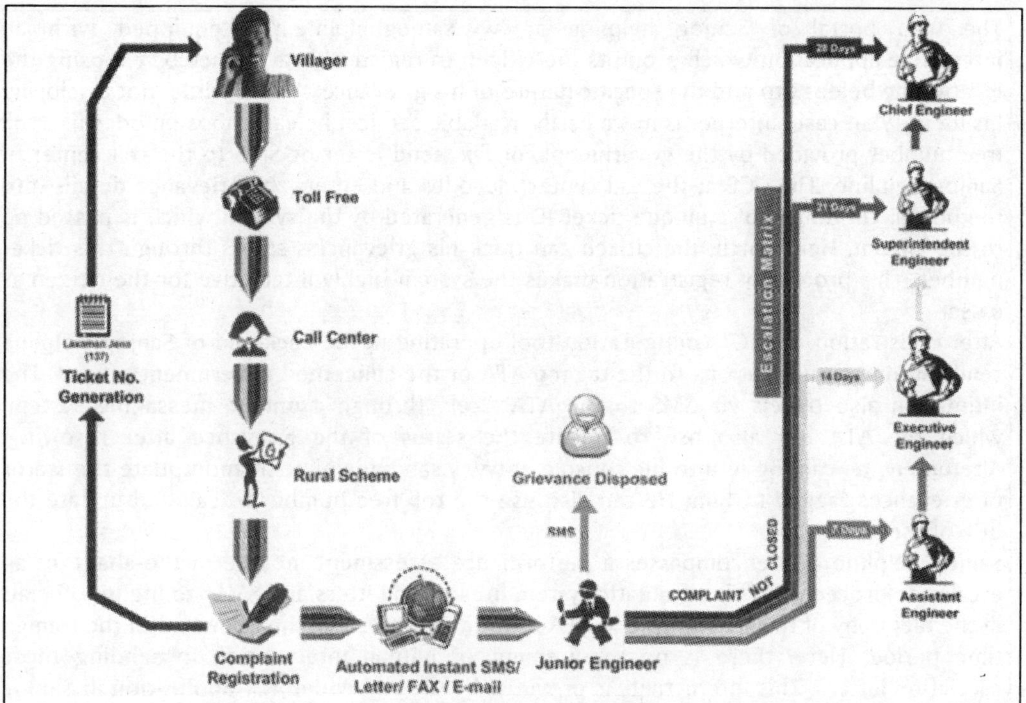

Delivering Citizen Centricity:

Sanjog helpline ensures a transparent grievance redressal service, for all the schemes in which the citizens are the immediate beneficiaries. Through this service, the citizen is empowered to register any type of complaint without disclosing their identity. This prevents apprehension by the citizens, ensuring confidentiality. They can register these complaints through 5 different communication channels namely - the web portal, fax, letter, SMS and toll free number. Having multi-channel communication ensures that the complaint is registered at Sanjog helpline positively. After registration, by a Call center executive (CCE), the citizens get a unique ticket number which helps them to track the status of their grievance anytime again with the help of the toll free number or the web portal.

The question however remains that how are the citizens benefited. The grievance registered at Sanjog Helpline, automatically gets forwarded to the concerned ATA. If the ATA is not able to resolve the complaint after checking its authenticity, within a stipulated time period, the complaint along with its details is forwarded to the higher officer of the scheme hierarchy. This mechanism of tracking the ATA by the system ensures faster intimation and disposal of the problem. The check-balance system of escalation ensures that the ATA is keen in resolving the issue within the specified time period. Therefore, the citizen can rest assured of his/her complaint getting resolved by the right person at the right time.

Sanjog Helpline's toll free number operates through its 9AM to 5PM Call Centre which translates the Grievance of a common citizen into the computer. The built in intelligent system ensures that the grievance is disposed by the use of ICT mechanism.

The web portal of Sanjog helpline at www.sanjoghelpline.in is equipped with an interactive application which prompts the citizen to register his grievance by choosing the location he belongs to and the specific nature of his grievance. All the while, not disclosing his identity! In case, Internet is not a easily available service, he can choose modes like toll free number provided by the government, or fax, send letter or SMS to the call center of Sanjog helpline. The CCE at the call center, decodes and enters the grievance details into the portal. On doing so, a unique ticket ID is generated by the system which is passed on to the citizen. Henceforth, the citizen can track his grievance's status through this ticker number. This process of registration makes the system highly interactive for the citizen as a user.

After registration, the ICT configuration tool operating in the back-end of Sanjog Helpline sends automatic intimations to the tagged ATA or the concerned government official. The intimation also travels via SMS to the ATA's cell, through a unique messaging system, which the ATA can also use to update the status of the grievance after resolving. Alternately, he can log in into his console at www.sanjoghelpline.in and update the status of grievances tagged to him. He can also use the toll free number to call and update the CCE of the status.

Sanjog helpline also encompasses a performance assessment module in the shape of an escalation procedure. This automatic system fires faxed letters and SMSs to higher officials in the hierarchy of the scheme, if the ATA is not able to resolve the issue within the framed time period. Here, there is no requirement of human intervention in sending these escalation letters. This interactivity is provided by the embedded ICT application in Sanjog helpline.

Impact Parameters and its productivity Analysis :

- **Political:** The usual rush of grievance seekers in the premises of the politically elected representatives has reduced tremendously as the citizens have been enlightened about the effectiveness of Sanjog helpline. Baral Munda of the Rimuli village has aptly put it. "With the advent of Sanjog-helpline we no longer have to queue in front of public or administrative representatives as they can directly complain at the Sanjog Helpline and wait patiently as they are assured that if their grievance is not resolved within a fixed period of time, it will be automatically escalated to the apex authority of the department or the organization."
- **Economical:** The citizen does not have to pay the transport fare to reach the grievance redressal Centre physically nor does he have pay any telephone charges if s/he wanted to make a call. S/he also is saved from paying bribes to the touts for the right contact and faster disposal.
- **Social:** The state of the art performance monitoring system that is built-in Sanjog Helpline helps the senior officers to measure and track the efficiency of their junior officers by monitoring the effective resolution of the registered grievances.
    - Communicate effectively across the schemes—Share operational performance management strategies throughout the govt. business enterprise.
    - Clarify business objectives—Align the effective implementation of welfare schemes with govt. strategy, plans and objectives

- Focus resources—Improve operational efficiency through a continuous business performance management cycle
- Increase govt. employee participation—Promote collaboration and accountability for results.
- Create value—Build future economic value for key govt. stakeholders as a single point of reference for centralized grievance repository while formulating welfare schemes in the future.

- **Technical:** Technical interventional has been deliberately kept as minimal as possible in order to make the solution more user-friendly and self-auditable. The technology is based on MVC (Model View Controller) architecture using a rule engine based pattern which isolates the business logic from the user interface, in such a scenario the user interface and the business logic can be independently configured & managed by the user.

  The Communication and the Data integration Interface Channels are based on service oriented methodology provided for system group functionality around business processes and are packaged as interoperable service. This methodology has been enabled to allow NREGASOFT software to exchange data with one another as they participate in business processes. Service orientation provides a loose coupling of services with the operating system, programming language and other technologies which under lie IGRS application. Use of relational database management system to facilitate data exchange between different departments.

- **Legal:** As it is know that the Govt. works with a Top to Bottom approach, where all instructions moving on an ink signed letter, this toll free grievance registration process comes as a boon. Hence it leads to a surge of complaints, resulting in huge communication flow and immense paper-work & never-ending follow ups.

The scanned Digital Signatures of the dignitaries are stored in the database and according to the intimation process set up in the system, the system will auto generate a reference numbered letter which will be dispatched in real time via fax, email besides being posted on the internet.

This fail safe methodology ascertains that the grievance is taken up seriously by the DPC besides being recorded in the govt. files. This in turn acts as a pressure tactics towards faster disposals.

The Govt. on its part legalized the whole process by gazetting the signature and the grievance compliance as mandatory and non-abeyance of deemed punishable by law.

- **Environmental:** This overall had a positive impact on the people as their faith in grass root governance which has been dwindling gradually because of lack of active feedback and participants of the benefactors, has been restored and their continuous patronage has helped the govt. to refine some its process wherever possible. This has proffered the public representative, the department and the citizenry a transparent and accountable self-evaluating and monitoring tool of universal appeal.

5.12 Conflict-An Introduction

Conflict is when two or more values, perspectives and opinions are contradictory in nature and haven't been aligned or agreed about yet, including:

1. Within yourself when you're not living according to your values;
2. When your values and perspectives are threatened; or
3. Discomfort from fear of the unknown or from lack of fulfillment. Conflict is inevitable and often good, for example, good teams always go through a "form, storm, norm and perform" period. Getting the most out of diversity means often-contradictory values, perspectives and opinions.

Conflict is often needed. It:

1. Helps to raise and address problems.
2. Energizes work to be on the most appropriate issues.
3. Helps people "be real", for example, it motivates them to participate.
4. Helps people learn how to recognize and benefit from their differences.

Conflict is not the same as discomfort. The conflict isn't the problem - it is when conflict is poorly managed that is the problem.

Conflict is a problem when it:

1. Hampers productivity.
2. Lowers morale.
3. Causes more and continued conflicts.
4. Causes inappropriate behaviors.

5.13 Phases of Conflict

- ***Prelude to Conflict***: Variables that make conflict possible between those involved
- ***Triggering Event***: A particular event, such as criticism which creates the conflict
- ***Initiation Phase***: Occurs when at least one person makes it known to the other that a conflict exists
- ***Differentiation Phase***: Parties raise the conflict issues and pursue reasons for the varying positions
- ***Integration stage/Resolution***: Parties acknowledge common grounds and explore possibilities to move towards a solution

5.14 Types of Conflict

A conceptual conflict can escalate into a verbal exchange and/or result in fighting. Conflict can exist at a variety of levels of analysis:

- community conflict
- diplomatic conflict
- economic conflict
- emotional conflict
- environmental resources conflict
- external conflict
- group conflict
- ideological conflict
- international conflict
- interpersonal conflict
- inter-social conflict
- intrastate conflict (for example: civil wars, election campaigns)
- intrapersonal conflict (though this usually just gets delegated out to psychology)
- organizational conflict
- intra-societal conflict
- military conflict
- Religious-based conflict (for example: Center For Reduction of Religious-Based Conflict).
- workplace conflict
- data conflict
- relationship conflict
- racial conflict

Conflicts in these levels may appear "nested" in conflicts residing at larger levels of analysis. For example, conflict within a work team may play out the dynamics of a broader conflict in the organization as a whole. (See Marie Dugan's article on Nested Conflict. John Paul Lederach has also written on this.) Theorists have claimed that parties can conceptualize responses to conflict according to a two-dimensional scheme; concern for one's own outcomes and concern for the outcomes of the other party. This scheme leads to the following hypotheses:

- High concern for both one's own and the other party's outcomes leads to attempts to find mutually beneficial solutions.
- High concern for one's own outcomes only leads to attempts to "win" the conflict.
- High concern for the other party's outcomes only leads to allowing the other to "win" the conflict.
- No concern for either side's outcomes leads to attempts to avoid the conflict.

Often a group finds itself in conflict over facts, goals, methods or values. It is critical that it properly identify the type of conflict it is experiencing if it hopes to manage the conflict through to resolution. For example, a group will often treat an assumption as a fact.

The more difficult type of conflict is when values are the root cause. It is more likely that a conflict over facts, or assumptions, will be resolved than one over values. It is extremely difficult to "prove" that a value is "right" or "correct". In some instances, a group will benefit from the use of a facilitator or process consultant to help identify the specific type of conflict. Practitioners of nonviolence have developed many practices to solve social and political conflicts without resorting to violence or coercion.

Conflict can arise between several characters and there can be more than one in a story or plot line. The little plot lines usually enhance the main conflict.

Conflict also defines as natural disagreement resulting from individuals or groups that differ in beliefs, attitudes, values or needs. It can also originate from past rivalries and personality differences. Other causes of conflict include trying to negotiate before the timing is right or before needed information is available. The following are the causes of conflict:

- communication failure
- personality conflict
- value differences
- goal differences
- methodological differences
- substandard performance
- lack of cooperation
- differences regarding authority
- differences regarding responsibility
- competition over resources
- non-compliance with rules (LO)

A definition of a conflict can be the subject of legal action has three invariants[1] :

- legal
- technical
- emotional

Types of Conflict

(1) Psychological Conflict (internal conflict)

- this type of conflict could be going on inside the person and no one would know (instinct may be at odds with values)
- Freud would say unconscious id battling superego, according to Freud our personalities are always in conflict

(2) Social Conflict

- interpersonal conflict- two individuals me against you; inter-group struggles -us against them; individual opposing a group- me against them, them against me; intragroup conflict- members of group all against each other on a task

(3) Approach-Avoidance

- conflict can be describe as having features of approach and avoidance: approach-approach conflict; avoidance-avoidance conflict; approach-avoidance conflict

Approach-Approach Conflict two desirable things are wanted, but only one option can be chosen (example: desirable date or ski trip)"I want this but I also want that."

Avoidance- avoidance Conflict two unattractive alternatives (example: study or do the dishes "I don't want his and I don't want that

Approach- Avoidance Conflict attractive and unattractive parts to both sides "I want this but I don't want what this entails"

(4) Functional vs. Dysfunctional Conflict

Dysfunctional Conflict: when conflict disrupts, hinders job performance, and upsets personal psychological functioning

Functional Conflict: from an interactionist perspective conflict can be responsive and innovative aiding in creativity and viability. Determine if conflict achieves goals or undermines them

Benefits of Conflict- Opportunity not obstacle by David Johnson

- enables a person to become aware of problems in a relationship
- serves as catalyst for positive change
- energizes and motivates to deal with immediate problems
- stimulates interest and curiosity
- relieves minor tensions -cathartic
- causes decision to be made more carefully
- promotes self-knowledge
- clears the air of unexpressed resentments

5.15 Types of Managerial Actions that Cause Workplace Conflicts

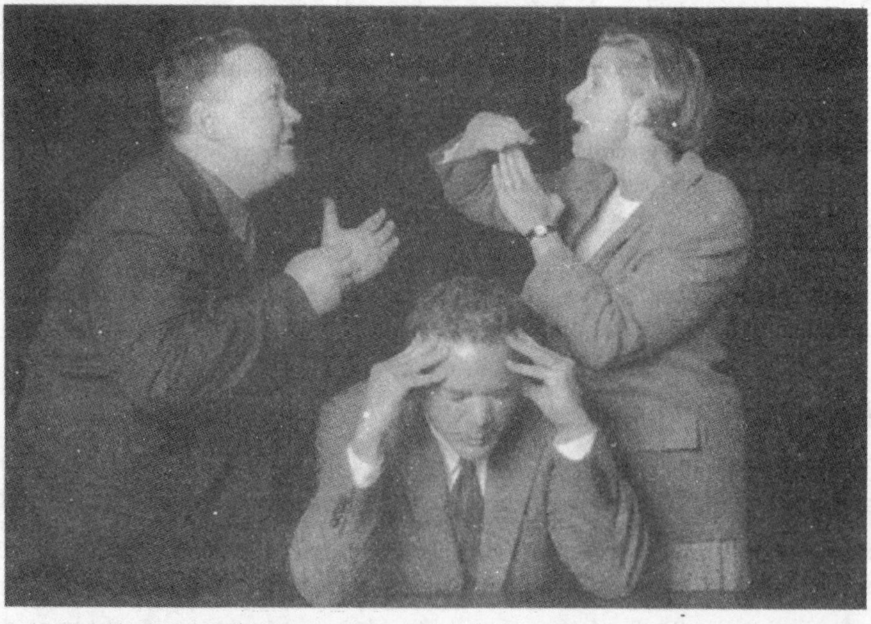

1. Poor communications

- Employees experience continuing surprises, they aren't informed of new decisions, programs, etc.

- Employees don't understand reasons for decisions, they aren't involved in decision-making.
- As a result, employees trust the "rumor mill" more than management.

2. The alignment or the amount of resources is insufficient. There is:
   - Disagreement about "who does what".
   - Stress from working with inadequate resources.
3. "Personal chemistry", including conflicting values or actions among managers and employees, for example:
   - Strong personal natures don't match.
   - We often don't like in others what we don't like in ourselves.
4. Leadership problems, including inconsistent, missing, too-strong or uninformed leadership (at any level in the organization), evidenced by:
   - Avoiding conflict, "passing the buck" with little follow-through on decisions.
   - Employees see the same continued issues in the workplace.
   - Supervisors don't understand the jobs of their subordinates.

5.16 Key Managerial Actions / Structures to Minimize Conflicts

- Regularly review job descriptions. Get your employee's input to them. Write down and date job descriptions. Ensure:
  - Job roles don't conflict.
  - No tasks "fall in a crack".

2. Intentionally build relationships with all subordinates.
    - Meet at least once a month alone with them in office.
    - Ask about accomplishments, challenges and issues.
3. Get regular, written status reports and include:
    - Accomplishments.
    - Currents issues and needs from management.
    - Plans for the upcoming period.
4. Conduct basic training about:
    - Interpersonal communications.
    - Conflict management.
    - Delegation.
5. Develop procedures for routine tasks and include the employees' input.
    - Have employees write procedures when possible and appropriate.
    - Get employees' review of the procedures.
    - Distribute the procedures.
    - Train employees about the procedures.
6. Regularly hold management meetings, for example, every month, to communicate new initiatives and status of current programs.
7. Consider an anonymous suggestion box in which employees can provide suggestions.

5.17 Ways to Deal with Conflict

There is no one best way to deal with conflict. It depends on the current situation. Here are the major ways that people use to deal with conflict.

1. Avoid it. Pretend it is not there or ignore it.
    - Use it when it simply is not worth the effort to argue. Usually this approach tends to worsen the conflict over time.
2. Accommodate it. Give in to others, sometimes to the extent that you compromise yourself.
    - Use this approach very sparingly and infrequently, for example, in situations when you know that you will have another more useful approach in the very near future. Usually this approach tends to worsen the conflict over time, and causes conflicts within yourself.

3. Competing. Work to get your way, rather than clarifying and addressing the issue. Competitors love accommodators.
   - Use when you have a very strong conviction about your position.
4 Compromising. Mutual give-and-take.
   - Use when the goal is to get past the issue and move on.
5. Collaborating. Focus on working together.
   - Use when the goal is to meet as many current needs as possible by using mutual resources. This approach sometimes raises new mutual needs.
   - Use when the goal is to cultivate ownership and commitment.

5.18 Management of Conflict within Yourself - "Core Process"

It's often in the trying that we find solace, not in getting the best solution. The following steps will help you in this regard.

1. Name the conflict, or identify the issue, including what you want that you aren't getting. Consider:
   - Writing your thoughts down to come to a conclusion.
   - Talk to someone, including asking them to help you summarize the conflict in 5 sentences or less.
2. Get perspective by discussing the issue with your friend or by putting it down in writing. Consider:
   - How important is this issue?
   - Does the issue seem worse because you're tired, angry at something else, etc.?
   - What's your role in this issue?
3. Pick at least one thing you can do about the conflict.
   - Identify at least three courses of action.
   - For each course, write at least three pros and cons.
   - Select an action - if there is no clear course of action, pick the alternative that will not hurt, or be least hurtful, to yourself and others.
   - Briefly discuss that course of action with a friend.
4. Then do something.
   - Wait at least a day before you do anything about the conflict. This

gives you a cooling off period.

- Then take an action.
- Have in your own mind, a date when you will act again if you see no clear improvement.

5.19 Management of Conflict with Another - "Core Process"

1. Know what you don't like about yourself, early on in your career. We often don't like in others what we don't want to see in ourselves.
   - Write down 5 traits that really bug you when see them in others.
   - Be aware that these traits are your "hot buttons".
2. Manage yourself. If you and/or the other person are getting heated up, then manage yourself to stay calm by
   - Speaking to the person as if the other person is not heated up - this can be very effective!
   - Avoid use of the word "you" - this avoids blaming.
   - Nod your head to assure them you heard them.
   - Maintain eye contact with them.
3. Move the discussion to a private area, if possible.
4. Give the other person time to vent.
   - Don't interrupt them or judge what they are saying.
5. Verify that you're accurately hearing each other. When they are done speaking:}
   - Ask the other person to let you rephrase (uninterrupted) what you are hearing from them to ensure you are hearing them.
   - To understand them more, ask open-ended questions. Avoid "why" questions - those questions often make people feel defensive.
6. Repeat the above step, this time for them to verify that they are hearing you. When you present your position
   - Use "I", not "you".
   - Talk in terms of the present as much as possible.
   - Mention your feelings.
7. Acknowledge where you disagree and where you agree.
8. Work the issue, not the person. When they are convinced that you understand them:

- Ask "What can we do fix the problem?" They will likely begin to complain again. Then ask the same question. Focus on actions they can do, too.

9. If possible, identify at least one action that can be done by one or both of you.
   - Ask the other person if they will support the action.
   - If they will not, then ask for a "cooling off period".
10. Thank the person for working with you.
11. If the situation remains a conflict, then:
   - Conclude if the other person's behavior conflicts with policies and procedures in the workplace and if so, present the issue to your supervisor.
   - Consider whether to agree to disagree.
   - Consider seeking a third party to mediate.

Summary

- A grievance is a written complaint, which can be used if employee thinks that the terms and condition of their job have changed unfairly or are not being honoured.
- A grievance is not a tool for dealing with suspensions, dismissals or competition appeals nor should this particular process be used to file harassment or discrimination of grievances.
- Grievances Handling Procedure is a method formulated by managerial authorities to deal with redressal of grievances.
- The grievances of the employees are related to the contract, work rule or regulation, policy or procedure, health and safety regulation etc.
- Conflict is when two or more values perspectives and opinions are contradictory in nature and haven't been aligned or agreed about yet.
- Phases of Conflict are –Prelude to conflict, Triggering Event, Initiation phase, Differentiation phase, Integration stage/Resolution.
- Dysfunctional conflict- when conflict disrupts, hinders job performance and upsets personal psychological function.
- Conflicts also defines as natural disagreement resulting from individuals or groups that differ in beliefs, attitudes , values or needs.

Objective type questions

1. Employees can express their dissatisfaction through
   (a) Grievances (b) Conflict
   (c) Industrial disputes (d) Suggestion boxes
2. Method formulated by managerial authorities to deal with redressal of grievances is -
   (a) Grievances Handling Procedure (b) Grievances redressal policy
   (c) Model Grievances policy (d) Conflict management
3. The Grievances handling procedure preferred by employees is -
   (a) Informal procedure (b) Open ended procedure
   (c) Single level procedure (d) Multi level procedure
4. Interpersonal conflict is related to -
   (a) Psychological Conflict (b) Social Conflict
   (c) Approach -Avoidance (d) Functional conflict
5. Natural disagreement resulting from individuals or groups that differ in beliefs, attitudes, values or need are-
   (a) Grievances (b) Conflict
   (c) Negligence (d) Industrial Dispute
6. Civil wars and election campaigns comes under-
   (a) Community Conflict (b) External Conflict
   (c) Intrastate Conflict (d) Ideological Conflict
7. Conflict which takes inside the person and is unknown to anyone is-
   (a) Community Conflict (b) External Conflict
   (c) Intra-state Conflict (d) Ideological Conflict
8. Workplace conflicts take place due to -
   (a) Lower morale (b) Poor communication
   (c) Productivity (d) Religious beliefs

Review Questions

1. Define Grievances.
2. What are the causes and effects of Grievances?

3. Mention the various types of Grievances?
4. What is Grievance handling procedure?
5. What are the various precautions and prescriptions involved in grievance handling?
6. How Grievances is managed in Indian industries?
7. Discuss in brief about grievance redressal policy.
8. Define conflict with its various phases.
9. What are the types of conflict and managerial actions that causes workplace conflict?
10. Mention various ways to deal with conflict?

Chapter 6

Industrial Disputes

6.1 Introduction

An Industrial dispute is defined as a difference and disagreement in opinion between two disputants generally between the worker and management or between employer and employees or between workmen and workmen, which is connected with the employment or non employment or the terms of employment or with the conditions of Labours or any person.

The scope of Industrial Dispute is wide. The word employment and non-employment in the definition of Industrial dispute are of widest amplitude and have been but in Just position to make the definition comprehensive. Any dispute concerned with employment or non-employment constitutes the subject matter of Industrial disputes.

According to section 2 (k) of Industrial Dispute Act, 1947, Industrial dispute is defined as:

"Any difference and disagreement between employees and employers or between employers and workmen or between workmen and workmen, which is connected with the employment or non-employment or the terms of employment or with conditions of Labours or any person"

6.2 Types of Industrial Disputes

There are two types of Industrial disputes– Interest disputes and Right disputes.

Interest dispute relates to the determination of new wage levels and other condition of employment while right disputes on the other hand relates to the interpretation and application of existing standards and usually involve and individual worker or group of workers under category of right disputes, claim is made that the workmen have not been treated in accordance with the rules, individual contract of employment, Laws and regulations any as per collective agreements. Such disputes are also described as grievance disputes. Such grievance may be regarding retrenchment, dismissal, payment of wages, working time, overtime, demotion, promotion, transfer, Seniority, job classification, work rules and fulfillment of obligation relating to safety and health laid down in an agreement.

6.3 Industrial Dispute Settlement

Settlement means a settlement arrived at in the course of conciliation proceeding and included a written agreement between employer and workmen arrived at otherwise than in course conciliation proceeding where such agreement has been signed by the parties thereto in such manner as may be prescribed and a copy thereof has been sent to the officer authorized in this behalf by the appropriate government and the conciliation officer. The definition envisages two categories of settlement.

(1) Settlement arrived at in the course of conciliation, and

(2) Settlement arrived at privately or otherwise than in the course of conciliation.

The settlement arrived at in the course of conciliation stand on a higher plane than the settlements arrived at otherwise than in the course of conciliation. The legal effect of both these settlements is not identical. The settlement arrived at otherwise than in the course conciliation binds only the parties to settlement and none else. In any case it does not stand on higher plane than the settlements arrived at in the conciliation and that makes the two distinct and different from each other.

Procedures for settling labour dispute: Collective Bargaining, Negotiation, Conciliation and Mediation, Arbitration and Adjudication are well known methods for settlement of industrial disputes.

(i) **Collective Bargaining**: - Collective Bargaining is a technique by which dispute as to conditions of employment, are resolved amicably, by agreement, rather than by coercion. The dispute is settled peacefully and voluntarily, although reluctantly, between labour and management. In the context of present day egalitarian society, with its fast changing social norms, a concept like 'collective bargaining' is not a capable of a precise definition. The content and Scope of collective bargaining also

varies from country to country. Broadly speaking Collective bargaining is a process of bargaining between employers and workers, by which they settle their disputes relating to employment or non-employment, terms of employment or conditions of the labour of the workman, among themselves, on the strength of the sanctions available to each side. Occasionally, such bargaining results in an amicable settlement, arrived at voluntarily and peacefully, between the parties. But quite often, the workers and employers have to apply sanctions by resorting to weapons of strike and lockouts, to pressurize one another, which makes both the sides aware of the strength of one another and that finally forces each of them to arrive at a settlement in mutual interest. It is thus the respective strength of the parties which determine the issue, rather than the wordy duals which are largely put on for show, as an element of strength in one party is by the same token, an element of weakness in another.

The final outcome of bargaining may also depend upon the art, skill and dexterity of displaying the strength by the representatives of one party to the other.

(ii) **Negotiation**: Negotiation is one of the principal means of settling labour disputes. However, due to lack of trust between the employers and workmen or their trade unions or inter-rivalry of the trade unions and the employers being in a commanding position, many a time negotiations fail. Through Amendment in the Act by Act 46 of 1982 Chapter II B providing for reference of certain individual disputes to Grievance Settlement Authority has been inserted in the Act. Under this Chapter, section 9 C has made it obligatory for the employers to make provision for Grievance Settlement Authority for settlement of industrial disputes connected with an individual workman employed in an establishment in which fifty or more workmen are employed or have been employed on any day. In the preceding twelve months. This amendment however even inspite of having been made twenty one years back has not seen the light of the day.

(iii) **Conciliation & Mediation:** Through conciliation and mediation a third party provides assistance with a view to help the parties to reach an agreement. The conciliator brings the rival parties together discuss with them their differences and assist them in finding out solution to their problems. Mediator on the other hand is more actively involved while assisting the parties to find an amicable settlement. Sometimes he submits his own proposals for settlement of their disputes.

Conciliation may be voluntary or compulsory. It is voluntary if the parties are free to make use of the same, while it is compulsory when the parties have to participate irrespective of whether they desire to do so or not. Section 4 of the Act provides for appointment for conciliation officers and Section 5 for constitution of Boards of Conciliation. The Board of conciliation is to consist of an independent Chairman and two or four member representing the parties in equal number. While the former is charged with the duty of mediating in and promoting the settlement of industrial disputes, the latter is required to promote the settlement of industrial disputes. The act generally allows registered trade unions or a substantial number of workers/ employees and also in certain cases individual workman to raise disputes. The performance of conciliation machinery, though it does not appear to be unsatisfactory, causes delays due to casual attitude of the parties towards conciliation, defective processes in the selection of personnel and unsatisfactory pre-job training and period-in-service-training. Delays in conciliation are attributed partly to the excessive work-load on officers and partly to the procedural defects. Since conciliation officer has no powers of coercion over labour and management , he can only persuade them to climb down and meet each other. The settlements that are claimed to result from conciliation are increasingly the result of political intervention. Success of conciliation depends upon the appearances and their sincere participation in conciliation proceedings of the parties before the conciliation officers. Non-appearance and non- participation of the parties in conciliation proceedings poses a serious hindrance in this direction. On the attitude of the parties National Commission on Labour observed conciliation is looked upon very often by the parties as merely hurdle to be crossed for reaching the next stage. The representatives sent by the parties to appear before him are generally officer who do not have the power to take decisions or make commitments: they merely carry the suggestion to the concerned authorities on either side. This dampens the spirit of a conciliator. We have been told by the employer's and workers, organizations alike that the conciliation machinery is weakened because of its falling into this type of disuse in recent years, Section 11 of the Act has clothed the conciliation officers with the power to enter premises occupied by any establishment and also has been invested with the powers of civil court under the Civil Procedure Code, 1908 when trying a suit for enforcing the attendance of any person and examining him on oath, compelling the production of documents and material objects and issuing commission for examination of witness for the purpose of inquiry in to any existing or apprehended industrial dispute. These provisions are seldom enforced. Moreover, conciliators most often do not have requisite information on the employers and trade unions, up to date wage/productivity, information and relevant up to date case laws which affect his capability to conciliate effectively. The National commission on labour in this context laid emphasis for pre job and on the job training of conciliation officers.

(iv) Arbitration and Adjudication

Arbitration: The resort to arbitration procedure may be compulsory or arbitrary. Compulsory arbitration is the submission of disputes to arbitration without consent or agreement of the parties involved in the dispute and the award given by the arbitrator being binding on the parties to the dispute. On the other hand in case of voluntary arbitration, the dispute can be referred for arbitration only if the parties agree to the same. Section 10 A of the Act, however, provides only for voluntary reference of dispute to arbitration. This system, however, has not been widely practiced so far. One of the main reasons for not gaining popularly of this procedure is lack of arbitrators who are able to command respect and confidence of the parties to the dispute. Inter Union rivalry also sometimes makes it difficult in arriving at an agreement on settlement of an arbitrator who is acceptable to all the trade unions in the industry.

The Apex court in case *Kurnal Leather Employess Union* vs *Liberty Footwear Co.* has held that the remedy under section 10 K is voluntary and alternative for settlement of industrial dispute but if the parties to the dispute have agreed in writing for settlement of their disputes through arbitrator, then the Govt. cannot refer the dispute to the Tribunal for adjudication.

Adjudication: If despite efforts of the conciliation officer, no settlement is arrived at between employer and the workman, The Industrial Dispute a provides for a three tier system of adjudication viz. Labour Courts , Industrial Tribunals and National Tribunals under section, 7, 7A and under section 7B respectively. Labour Courts have been empowered to decide disputes relating to matters specified in the Second Schedule. These matters are concerned with the rights of workers, such as propriety of legality of an order passed by an employer under the standing orders, application and interpretation of standing orders, discharge or dismissal of workman including reinstatement of grant of relief to workman wrongfully discharged or dismissed, withdrawal of any customary concession or privilege and illegality or otherwise of a strike or lockout.

The industrial tribunals are empowered to adjudicate on matters specified in both the Second and third schedule i.e. both rights and interest disputes. The jurisdiction of the Industrial Tribunal is wider that the labour courts.

In Paulos vs State of Kerala, per Mathew J. The government entrusted the work of selection of candidates for the appointment of presiding officers of industrial tribunals and labour courts to the advocate-general. This mode of selection of candidates was challenged by writ petition in High Court of Kerala on the ground that the government is bound to make appointment to this post after giving on opportunity to all eligible persons before considering for appointment by proper publicity through advertisement in newspapers. In the absence of such opportunity being given to all the persons having such prescribed qualification to be appointed, the method was unfair and arbitrary, and, therefore, violation of Articles 14 and 16 of the Constitution. A Single Judge of the Kerala High Court upheld the appointment holding that the action taken by the government was within the powers enjoined by law and it is not the requirement of law that for every recruitment to an office under state, there must be an advertisement in the public press. Therefore, it is not necessary that the state must in every case of public employment issue an advertisement or notice inviting applications for an office.

In *Shellac Industries Ltd* vs *Workmen,* per Dutt J. A tribunal once appointed cannot be abolished by an executive act merely because the government chooses to put an end to it when a reference is pending before it, for the state cannot do indirectly what is not permissible to it to do expressly or impliedly under the Act. Hence, a dispute pending before such a tribunal cannot be referred to another tribunal under Sec 10 (1) (d) as that can be done only under Section 33 B.

In case of disputes which in the opinion of the Central Govt. involves question of national importance or is of such nature that workers in more than one State are likely to be affected. The Act provides for constitution of National Tribunals.

Industrial adjudication has undoubtedly played a conclusive role in the settlement of industrial disputes and in ameliorating the working and living conditions of labour class. In this context the National Commission of Labour observed: the adjudicating machinery has exercised considerable influence on several aspects of conditions of work and labour management relations. Adjudication has been on of the instruments for the improvement of wages and working conditions and for securing allowances for maintaining real wages, bonus and introducing uniformity in benefits and amenities. It has also helped to avert many work stoppages by providing an acceptable alternative to direct action and to protect and promote the interest of the weaker sections of the working class, who were not well organized or were unable to bargain on an equal footing with the employer.

The Act empowers the appropriate government to refer industrial disputes when the industrial disputes exist or are apprehended. The Apex court has also held in *Shambu Nath* vs *Bank of Baroda* that the power conferred by Section 10 (1) on the Govt. to make reference can be exercised not only when an industrial dispute exists but when it is also apprehended. *Kotwal J. Kashmir Ceramtics Ltd.* vs *Labour Court* It is not permissible for the labour court to entertain more disputes than are contemplated in the reference not is it permissible for it to decline to adjudicate matters which clearly arise in the terms of the reference.

In the case *State of Madras* vs *C.P. Sarathi* and *Secretary, India Tea Association* vs *Ajeet Kumar Bharat*, it was held that to make a reference is the administrative act of the Government and the same view has been taken in the case *Telecom Conway Divers Mazdoor Sangh* vs *State of Bihar* and in *M/s Avon Services (Production Agencies) Pvt. Ltd* vs *Industrial Tribunal Faridabad* with the result that the State Government has little choice in referring to make references of the disputes after failure of conciliation proceedings . The adjudication system is not immune from its weakness. The adjudication is dilatory and expensive. The Apex Court in case *Ajaib Singh* vs *Sirhind Co Op. Marketing Cum Processing Service Society Ltd* has also held that reference of industrial dispute to labour court is not subject to limitation under Article 137 of the limitation Act . Thus no period of limitation having been prescribed under the Act during which the industrial disputes can be raised and referred for adjudication sometimes state disputes which arose even 15 to 20 years back are referred for adjudication. Moreover the Labour court , Tribunal and National Tribunal do not posses power of executing the order/awards passed by them although they are presided over by highly qualified and experienced judicial officers such as District Judges and High Court Judges with the result that generally workmen, weaker sections of the society suffer on account of non-implementation of the order/awards. However, there is no viable alternative to this system. Stringent provisions, therefore are required for ensuring the time limit within which the orders /awards to be implemented and clothe the courts and tribunal with powers of contempt of court for non-implementation of orders /awards passed by them.

Under the Act, an award made by the adjudication authority is final as there is no appeal. However actual practice almost every award made against the employer is challenged in the High Court under Articles 226 and 227 & in the Supreme Court under Article 136. It takes year before final orders are passed in writ petitions pending before the High Court/Supreme Court. If the period taken before the adjudicating authority is counted, it does not take less 10 to 20 years before the protracted litigation could be disposed off. It is the weaker sections who are inconvenienced and handicapped the most, by the delay.

It is submitted that the need of the day is to evolve the frame-work in which workers and the management perceive the need to co-operate. Bilateral regulation is the most effective method of evolving norms which enjoy wide acceptance.

6.4 Strikes and Lockouts

Strike is defined as a cessation of work by a body of persons employed in any industry acting in combination or a concerted refusal or a refusal under a common under standing of any number of persons who are or have been so employed to continue to work or to accept employment. A lock out is a temporary closing of a place of employment or the suspension of work or the refusal by an employer to continue to employ any number of persons employed by him.

The basic unit of measurement used to record a strike or lockouts is the case of dispute. A strike or lockout that is interrupted and later resumes, still due to the same case of dispute, is treated as a new strike or lockout when it resumes. If work stoppages due to the same case of dispute occur simultaneously in different establishments of the same enterprise they are considered to be one strike or lockout. Similarly, if stoppages, due to the same case of dispute, occur simultaneously in establishments of different enterprises, they are treated as one strike or lockout; parameters such as industry, sector of ownership, result, method of settlement and affiliation to the central worker's or employers organization are also kept in mind when deciding whether the stoppage is a single or lockout. Stoppages due to the same case of dispute occurring at different times are treated as different strike or lockouts.

6.5 Types of Strikes

Some important strike's which generally flourishes in any industry are mentioned below:

1. **General Strike**: In this strike, all or most of the members of union in a location or in any industry, organizes strike for the demands of all workers. This strike is also referred as extension of sympathetic strike which expresses a generalized protest by the workers.

2. **Sympathetic Strike**: In this case, strikes are organised on the basis of sympathy i.e. if workers of one division are under strike then the workers of another division also moves on strike for the sake of sympathy towards the workers of further one division.

3. **Economic Strike**: In this case, strikes are organised on the basis of increasing the economic standards of worker i.e. Demand of increase in wages, allowances like rent allowance, dearness allowance, travelling allowance and bonus, by the workers results in Economic strike.

4. **Stay in strike:** In this strike, workers neither work nor become absent, but control the production facility, i.e. workers are on strike, while they are present and they doesn't perform any production work. This strike is also known as tool down strike or Pen down strike.

5. **Slow down strike:** In this case, workers organize strike, by remaining on their job and slowing down their work action. This act is called as Go-slow tactics. This restricted rate of output by workers imposes heavy pressure on employers and may results in stoppage of work completely.

6. **Sick out-in:** In this case, group of workers takes sickness leave for a particular period. This mass sickness leave results in sudden loss and creates an atmosphere of strike.

7. **Boycott:** In this case, workers boycott the products of company and makes appeal generally in public.

8. **Picketing:** In this case, workers are dissuaded from work by certain workers who were standing at the gate of factory. This strike system is legal until any violence takes place.

9. **Gherao:** In this case, the managerial authorities were blocked from getting outside and forced them to stay inside their cabins. This action imparts a physical and mental torture to the managerial authorities.

6.6 Case study: Coal Strike of 1902

The **Coal Strike of 1902** was a strike by the United Mine Workers of America in the Anthracite Coal Fields of eastern Pennsylvania. The strike threatened to shut down the winter fuel supply to all major cities (homes and apartments were heated with anthracite or "hard" coal because it had higher heat value and less smoke than "soft" or bituminous coal). President Theodore Roosevelt became involved and set up a fact-finding commission that suspended the strike. The strike never resumed, as the miners received more pay for fewer hours; the owners got a higher price for coal, and did not recognize the union as a bargaining agent. It was the first labor episode in which the federal government intervened as a neutral arbitrator.

1. The 1899 and 1900 strikes

The UMWA had won a sweeping victory in the 1897 strike by the soft-coal (bituminous coal) miners in the Midwest, winning significant wage increases and growing from 10,000 to 115,000. A number of small strikes took place in the anthracite district from 1899 to 1901, by which the union gained experience and unionized more workers. The 1899 strike in Nanticoke, Pennsylvania, demonstrated that the unions could win a strike directed against a subsidiary of one of the large railroads. [Blatz 1991] It hoped to make similar gains in 1900, but found the operators, who had established an oligopoly through concentration of ownership after drastic fluctuations in the market for anthracite, to be far more determined opponents than it had anticipated. The owners refused to meet or to arbitrate with the union; the union struck on September 17, 1900, with results that surprised even the union, as miners of all different nationalities walked out in support of the union.

Coal miners in Hazleton, PA. 1905

Republican Senator Mark Hanna, himself an owner of bituminous coal mines (not involved in the strike) sought to resolve the strike, coming less than two months before the presidential election. He worked through the National Civil Federation which brought labor and capital together. Relying on J. P. Morgan to convey his message to the industry that a strike would hurt the reelection of Republican William McKinley, Hanna was able to convince the owners to concede a wage increase and grievance procedure to the strikers. The industry refused, on the other hand, to formally recognize the UMWA as the representative of the workers. The union declared victory and dropped its demand for union recognition.

2. The anthracite coal strike of 1902

The issues that lead to the strike of 1900 were just as pressing in 1902: the union wanted recognition and a degree of control over the industry. The industry, still smarting from its concessions in 1900, opposed any federal role. The 150,000 miners wanted their weekly pay envelope. Tens of millions of city dwellers needed coal to heat their homes.

John Mitchell, President of the UMWA, proposed mediation through the National Civic Federation, then a body of relatively progressive employers committed to collective as a means of resolving labor disputes. In the alternative, Mitchell proposed that a committee of eminent clergymen report on conditions in the coalfields. George Baer, President of the Philadelphia and Reading Railroad, one of the leading employers in the industry, brushed aside both proposals dismissively:

Miner strikes the owner (1902 *Judge* Cartoon)

"Anthracite mining is a business and not a religious, sentimental, or academic proposition.... I could not if I would delegate this business management to even so highly a respectable body as the Civic Federation, nor can I call to my aid . . . the eminent prelates you have named."

On May 12, 1902, the miners went out on strike. The maintenance employees had much steadier jobs and did not face the special dangers of underground work, walked out on June 2. The union had the support of roughly eighty percent of the workers in this area, or more than 100,000 strikers. Some 30,000 left the region, many headed for Midwestern bituminous mines; 10,000 returned to Europe. The strike soon produced threats of violence between the strikers on one side and strikebreakers, the Pennsylvania National Guard, local police and hired detective agencies on the other.

3. Federal intervention

On June 8 President Theodore Roosevelt asked his Commissioner of Labor, Carroll D. Wright, to investigate the strike. Wright investigated and proposed reforms that acknowledged each side's position, recommending a nine hour day on an experimental basis and limited collective bargaining. Roosevelt chose not to release the report, for fear of appearing to side with the union.

The owners, for their part, refused to negotiate with the union. As George Baer wrote when urged to make concessions to the strikers and their union, the "rights and interests of the laboring man will be protected and cared for—not by the labor agitators, but by the Christian men to whom God in His infinite wisdom has given the control of the property interests of the country." The union used this letter to sway public opinion behind the strike.

Roosevelt wanted to intervene, but he was told by his Attorney General, Philander Knox, that he had no authority to do so. Mark Hanna and many others in the Republican Party were likewise concerned about the political implications if the strike dragged on into winter, when the need for anthracite was greatest. As Roosevelt told Hanna, "A coal famine in the winter is an ugly thing and I fear we shall see terrible suffering and grave disaster."

WHAT DIDN'T HAPPEN IN THE COAL CONTROVERSY—*Cleveland Plain Dealer*

The two sides were supposed to listen to expert testimony and come to a friendly agreement; cartoon from the Cleveland Dealer

Roosevelt therefore convened a conference of representatives of government, labor, and management on October 3, 1902. The union considered the mere holding of a meeting to be tantamount to union recognition and took a conciliatory tone. The owners told Roosevelt that strikers had killed over 20 men and that he should use the power of government "to protect the man who wants to work, and his wife and children when at work." With proper protection they would produce enough coal to end the fuel shortage; they refused to enter into any negotiations with the union. The governor sent in the National Guard, who protected the mines and the minority of men still working.

Roosevelt attempted to persuade the union to end the strike with a promise that he would create a commission to study the causes of the strike and propose a solution, which Roosevelt promised to support with all of the authority of his office. Mitchell refused and his membership endorsed his decision by a nearly unanimous vote.

The economics of coal revolved around two factors: most of the cost of production was wages for miners, and if the supply fell the price would shoot up because in an age before oil and electricity, there were no good substitutes. Profits were low in 1902 because of an over supply; therefore the owners welcomed a moderately long strike. They had huge stockpiles which grew in value daily. It was illegal for the owners to conspire to shut down production, but not so if the miners went on strike. The owners welcomed the strike, but they adamantly refused to recognize the union, for then it would control the coal industry.

THE WASHINGTON SCHOOLMASTER
From the *Chronicle* (Chicago)

TR teaches the childish coal barons a lesson; 1902 editorial cartoon

Roosevelt continued to try to build support for a mediated solution, persuading former President Grover Cleveland to join the commission he was creating. He also considered sending the U.S. Army to take over the coalfields.

4. J.P. Morgan intervenes

J.P. Morgan, the dominant figure in American finance, had played a role in resolving the 1900 strike. He was deeply involved in this strike as well: his interests included the Reading Railroad, one of the largest employers of miners, and he had installed George Baer, who spoke for the industry throughout the strike, as the head of the railroad.

Now, at the urging of Secretary of War, Elihu Root, Morgan came up with another compromise proposal that provide for arbitration, while giving the industry the right to deny that it was bargaining with the union by directing that each employer and its employees communicate directly with the commission. The employers agreed on the condition that the five members be a military engineer, a mining engineer, a judge, an expert in the coal business, and an "eminent sociologist". The employers were willing to accept a union leader as the "eminent sociologist," so Roosevelt named E. E. Clark, head of the railway conductors' union, as the "eminent sociologist" and, after Catholics exerted pressure, added a sixth, Catholic bishop John Lancaster Spaulding, and Commissioner Wright as the seventh member.

5. The Anthracite Coal Strike Commission

The anthracite strike ended, after 163 days, on October 23, 1902. The commissioners began work the next day, then spent a week touring the coal regions. Wright used the staff of the Department of Labor to collect data about the cost of living in the coalfields.

The commissioners then held hearings over the next three months, taking testimony from 558 witnesses, including 240 for the striking miners, 153 for nonunion mineworkers, 154 for the operators and eleven called by the Commission itself. Baer made the closing arguments for the coal operators, while lawyer Clarence Darrow closed for the workers.

Although the commissioners heard some evidence of terrible conditions, they concluded that the "moving spectacle of horrors" represented only a small number of cases. By and large, social conditions in mine communities were found to be good, and miners were judged as only partly justified in their claim that annual earnings were not sufficient "to maintain an American standard of living."

Baer said in his closing arguments, "These men don't suffer. Why, hell, half of them don't even speak English". Darrow, for his part, summed up the pages of testimony of mistreatment he had obtained in the soaring rhetoric for which he was famous: "We are working for democracy, for humanity, for the future, for the day will come too late for us to see it or know it or receive its benefits, but which will come, and will remember our struggles, our triumphs, our defeats, and the words which we speak."

In the end, however, the rhetoric of both sides made little difference to the Commission, which split the difference between mineworkers and mine owners. The miners asked for 20% wage increases, and most were given a 10% increase. The miners had asked for an eight-hour day and were awarded a nine-hour day instead of the standard ten hours then prevailing. While the operators refused to recognize the United Mine Workers, they were required to agree to a six-man arbitration board, made up of equal numbers of labor and management representatives, with the power to settle labor disputes. Mitchell considered that de facto recognition and called it a victory.

6. The aftermath of the strike

Organized labor celebrated the outcome as a victory for the UMWA and unions generally. Membership in other union's soared, and moderate leaders gained more than the radical Socialists who denounced capitalism but were forced to the fringe of organized labor. Young John Mitchell proved his leadership skills and mastership of the problems of ethnic, skill, and regional divisions that had long plagued the union in the anthracite region. By contrast the Western Federation of Miners' strikes in the West often turned into full-scale warfare between strikers and both employers and the civil and military authorities. This strike was successfully mediated through the intervention of the federal government, which strove to provide a "Square Deal"—which Roosevelt took as the motto for his administration—to both sides. The settlement was an important step in the Progressive reforms of the decade that followed. There were no more major coal strikes until the 1920s.

6.7 The Industrial Disputes Act, 1947 (11th March, 1947)

This was an act to make provision for the investigation and settlement of industrial disputes, and for certain other purposes.

This act consist of

Chapter I-Preliminary

1. Short title, extent and commencement:
    - this act may be called the Industrial Disputes Act, 1947.
    - it extends to the whole of India.
    - it shall into force on the first day of April, 1947.

2. **Definitions:-** It consists of definitions of terms like:

- **Appropriate Government:-** It may be Central or State Governments which empowers authority on various industries.
- **Average pay:-** It means the average of the wages payable to a workman in following situations:

  (i) in the case of monthly paid workman, in the three complete calendar on the

  (ii) in the case of weekly paid workman, in the four complete weeks

  (iii) in the case of daily paid workman, in the twelve full working days
- **Award:** It means an interims or a final determination of any industrial dispute or of any questions relating there to by any Labour court, Industrial tribunal or National Industrial tribunal.
- **Board:** It means a Board of conciliation constituted under this act.
- **Conciliation proceeding:** It means a conciliation officer appointed under this Act.
- **Conciliation proceeding:** It means any proceeding held by a conciliation officer or Board under this Act.
- **Controlled industry:** It means any industry the control of which by the union has been declared by any central Act to be expedient in the public interest.
- **Court:** It means a court of Inquiry constituted under this Act.
- **Lock out:** It means the temporary closing of a place of employment or the suspension of work, or the refusal by an employer to continue to employ any number of persons employed by him.
- **Retrenchment:** It means the termination by the employer of the service of a workman for any reason what so ever, otherwise than as a punishment inflicted by way of disciplinary action, but does not include:

  (i) voluntary retirement of the workman

  (ii) retirement of the workman on reaching the age of super annulations if the contract of employment between the employer and the Workman concerned contains a stipulation in that behalf.

  (iii) termination of the service of a workman on the ground of continued ill health.
- **Settlement:** It means a settlement arrived at in the course of conciliation proceeding and includes a written agreement between the employer and

workman arrived at otherwise than in the course of conciliation proceeding where such agreement has been signed by the parties there to in such manner as many be prescribed and a copy there of has been sent to appropriate Government and the conciliation officer.

- **Strike:** It means a cessation of work by a body of persons employed in any industry acting in combination or a concerted refusal, or a refusal under a common understanding, of any number of persons who are or have been so employed to continue to work or to accept employment.
- **Wages:** It means all remuneration capable of being expressed in terms of money, which would, if the terms of employment, expressed or implied, were fulfilled be payable to a workman in respect of his employment or of work done in such employment and includes.

  (i) Such allowances (including dearness allowances) as the workman is for the time being entitled to.

  (ii) the value of any house accommodation, or of supply of light, water, medical attendance or other amenity or of any service or of any concessional supply of food-grains or other articles.

  (iii) any traveling concession

  (iv) any commission payable on the promotion of sales or business or both.

But it does not includes:-

(i) any bonus

(ii) any contribution paid or payable by the employer to any pension fund or provident fund or for the benefits of the workman under any Law for the time being in force.

(iii) any gratuity payable on the termination of his service

- **Workman:** It means any person (including an apprentice) employed in any industry to do any manual, unskilled, skilled, technical, operational, clerical or super visory work for hire or reward, whether the terms of employment be express or implied, and for the purpose of any proceeding under this act in relation to an industrial dispute, includes any such person who has been dismissed, discharged or retrenched in connection with, or as a consequence of that dispute or whose dismissal, discharge or retrenchement has led to that dispute, but does not include any such person-

  (i) who is subject to the Air Force Act, 1950 (45 of 1950), or the Army Act, 1950 (46 of 1950), or the Navy Act, 1957 (62 of 1957); or

(ii) who is employed in the police service or as an officer or other employee of a prison; or

(iii) who is employed mainly in a managerial or administrative capacity; or

(iv) who, being employed in a supervisory capacity, draws wages exceeding one thousand six hundred rupees per mensem or exercises, either by the nature of the duties attached to the office or by reason of the powers vested in him, functions mainly of a managerial nature.

CHAPTER II -AUTHORITIES UNDER THIS ACT

1. Works Committee.

Features are :-

(1) In the case of any industrial establishment in which one hundred or more workmen are employed or have been employed on any day in the preceding twelve months, the appropriate Government may by general or special order require the employer to constitute in the prescribed manner a Works Committee consisting of representatives of employers and workmen engaged in the establishment so however that the number of representatives of workmen on the Committee shall not be less than the number of representatives of the employer. The representatives of the workmen shall be chosen in the prescribed manner from among the workmen engaged in the establishment and in consultation with their trade union, if any, registered under the Indian Trade Unions Act, 1926 (16 of 1926).

(2) It shall be the duty of the Works Committee to promote measures for securing and preserving amity and good relations between the employer and workmen and, to that end, to comment upon matters of their common interest or concern and endeavour to compose any material difference of opinion in respect of such matters.

2. Conciliation officers.

Features are :-

(1) The appropriate Government may, by notification in the Official Gazette, appoint such number of persons as it thinks fit, to be conciliation officers, charged with the duty of mediating in and promoting the settlement of industrial disputes.

(2) A conciliation officer may be appointed for a specified area or for specified industries in a specified area or for one or more specified industries and either permanently or for a limited period.

3. Boards of Conciliation.

Features are :-

(1) The appropriate Government may as occasion arises by notification in the Official Gazette constitute a Board of Conciliation for promoting the settlement of an industrial dispute.

(2) A Board shall consist of a chairman and two or four other members, as the appropriate Government thinks fit.

(3) The chairman shall be an independent person and the other members shall be persons appointed in equal numbers to represent the parties to the dispute and any person appointed to represent a party shall be appointed on the recommendation of that party:

Provided that, if any party fails to make a recommendation as aforesaid within the prescribed time, the appropriate Government shall appoint such persons as it thinks fit to represent that party.

(4) A Board, having the prescribed quorum, may act notwithstanding the absence of the chairman or any of its members or any vacancy in its number:

Provided that if the appropriate Government notifies the Board that the services of the chairman or of any other member have ceased to be available, the Board shall not act until a new chairman or member, as the case may be, has been appointed.

4. Courts of Inquiry.

Features are :-

(1) The appropriate Government may as occasion arises by notification in the Official Gazette constitute a Court of Inquiry for inquiring into any matter appearing to be connected with or relevant to an industrial dispute.

(2) A Court may consist of one independent person or of such number of independent persons as the appropriate Government may think fit and where a Court consists of two or more members, one of them shall be appointed as the chairman.

(3) A Court, having the prescribed quorum, may act notwithstanding the absence of the chairman or any of its members or any vacancy in its number:

Provided that, if the appropriate Government notifies the Court that the services of the chairman have ceased to be available, the Court shall not act until a new chairman has been appointed.

5. Labour Courts.

Features are :-

(1) The appropriate Government may, by notification in the Official Gazette, constitute one or more Labour Courts for the adjudication of industrial disputes relating to any matter specified in the Second Schedule and for performing such other functions as may be assigned to them under this Act.

(2) A Labour Court shall consist of one person only to be appointed by the appropriate Government.

(3) A person shall not be qualified for appointment as the presiding officer of a Labour Court, unless-

(a) he is, or has been, a Judge of a High Court; or

(b) he has, for a period of not less than three years, been a District Judge or an Additional District Judge; or

(c) he has held any judicial office in India for not less than seven years; or

(d) he has been the presiding officer of a Labour Court constituted under any Provincial Act or State Act for not less than five years.

6. Tribunals.

Features are :-

(1) The appropriate Government may, by notification in the Official Gazette, constitute one or more Industrial Tribunals for the adjudication of industrial disputes relating to any matter, whether specified in the Second Schedule or the Third Schedule.

(2) A Tribunal shall consist of one person only to be appointed by the appropriate Government.

(3) A person shall not be qualified for appointment as the presiding officer of a Tribunal unless -

(a) he is, or has been, a Judge of a High court or he has, for a period of not less than three years, been a district judge or an Additional District Judge.

(4) The appropriate Government may, if it so thinks fit, appoint two persons as assessors to advise the Tribunal in the proceeding before it.

7. National Tribunals.-

(1) The Central Government may, by notification in the Official Gazette, constitute one or more National Industrial Tribunals for the adjudication of industrial disputes which, in the opinion of the Central Government, involve questions of

national importance or are of such a nature that industrial establishments situated in more than one State are likely to be interested in, or affected by, such disputes.

(2) A National Tribunal shall consist of one person only to be appointed by the Central Government.

(3) A person shall not be qualified for appointment as the presiding officer of a National Tribunal.

(4) The Central Government may, if it so thinks fit, appoint two persons as assessors to advise the National Tribunal in the proceeding before it.

Disqualifications for the presiding officers of Labour Courts,Tribunals and National Tribunals.

Disqualifications for the presiding officers of Labour Courts, Tribunals and National Tribunals.-No person shall be appointed to, or continue in, the office of the presiding officer of a Labour Court, Tribunal or National Tribunal, if--

(a) he is not an independent person; or

(b) he has attained the age of sixty-five years.]

8. Filling of vacancies.-

If, for any reason a vacancy (other than a temporary absence) occurs in the office of the presiding officer of a Labour Court, Tribunal or National Tribunal or in the office of the chairman or any other member of a Board or Court, then, in the case of a National Tribunal, the Central Government and in any other case, the appropriate Government, shall appoint another person in accordance with the provisions of this Act to fill the vacancy, and the proceeding may be continued before the Labour Court, Tribunal, National Tribunal, Board or Court, as the case may be, from the stage at which the vacancy is filled.

9. Finality of orders constituting Boards

(1) No order of the appropriate Government or of the Central Government appointing any person as the chairman or any other member of a Board or Court or as the presiding officer of a Labour Court, Tribunal or National Tribunal shall be called in question in any manner; and no act or proceeding before any Board or Court shall be called in question in any manner on the ground merely of the existence of any vacancy in, or defect in the constitution of, such Board or Court.

(2) No settlement arrived at in the course of a conciliation proceeding shall be invalid by reason only of the fact that such settlement was arrived at after the expiry of the period referred to in sub-section (6) of section 12 or sub-section (5) of section 13, as the case may be.

(3) Where the report of any settlement arrived at in the course of conciliation proceeding before a Board is signed by the chairman and all the other members of the Board, no such settlement shall be invalid by reason only of the casual or unforeseen absence of any of the members (including the chairman) of the Board during any stage of the proceeding.]

CHAPTER III

REFERENCE OF DISPUTES TO BOARDS, COURTS OR TRIBUNALS

10. Reference of disputes to Boards, Courts or Tribunals.-

(a) Where the appropriate Government is of opinion that any industrial dispute exists or is apprehended, it may at any time], by order in writing,--

- refer the dispute to a Board for promoting a settlement thereof; or
- refer any matter appearing to be connected with or relevant to the dispute to a Court for inquiry; or
- refer the dispute or any matter appearing to be connected with, or relevant to, the dispute, if it relates to any matter specified in the Second Schedule, to a Labour Court for adjudication; or
- refer the dispute or any matter appearing to be connected with, or relevant to, the dispute, whether it relates to any matter specified, in the Second Schedule or the Third Schedule, to a Tribunal for adjudication:
- Provided that where the dispute relates to any matter specified in the Third Schedule and is not likely to affect more than one hundred workmen, the appropriate Government may, if it so thinks fit, make the reference to a Labour Court under clause (c)

(b) [Provided further that] where the dispute relates to a public utility service and a notice under section 22 has been given, the appropriate Government shall, unless it considers that the notice has been frivolously or vexatiously given or that it would be inexpedient so to do, make a reference under this subsection notwithstanding that any other proceedings under this Act in respect of the dispute may have commenced:

(c) [Provided also that where the dispute in relation to which the Central Government is the appropriate Government, it shall be competent for that Government to refer the dispute to a Labour Court or an Industrial Tribunal, as the case may be, constituted by the State Government;]

(d) Where the Central Government is of opinion that any industrial dispute exists or is apprehended and the dispute involves any question of national importance or is of such a nature that industrial establishments situated in more than one State are likely to be interested in, or affected by, such dispute and that the dispute should be adjudicated by a National Tribunal, then, the Central Government may, whether or not it is the appropriate Government in relation to that dispute, at any time, by order in writing, refer the dispute or any matter appearing to be connected with, or relevant to, the dispute, whether it relates to any matter specified in the Second Schedule or the Third Schedule, to a National Tribunal for adjudication.

(2) Where the parties to an industrial dispute apply in the prescribed manner, whether jointly or separately, for a reference of the dispute to a Board, Court [Labour Court, Tribunal or National Tribunal], the appropriate Government, if satisfied that the persons applying represent the majority of each party, shall make the reference accordingly.

- An order referring an industrial dispute to a Labour Court, Tribunal or National Tribunal under this section shall specify the period within which such Labour Court, Tribunal or National Tribunal shall submit its award on such dispute to the appropriate Government:

(3) Where an industrial dispute has been referred to a Board, [Labour Court, Tribunal or National Tribunal] under this section, the appropriate Government may by order prohibit the continuance of any strike or lock-out in connection with such dispute which may be in existence on the date of the reference.

(4) Where in an order referring an industrial dispute to [a Labour Court, Tribunal or National Tribunal] under this section or in a subsequent order, the appropriate Government has specified the points of dispute for adjudication, [the Labour Court or the Tribunal or the National Tribunal, as the case may be] shall confine its adjudication to those points and matters incidental thereto.

(5) Where a dispute concerning any establishment or establishments has been, or is to be, referred to a [Labour Court, Tribunal or National Tribunal] under this section and the appropriate Government is of opinion, whether on an application made to it in this behalf or otherwise, that the dispute is of such a nature that any other establishment, group or class of establishments of a similar nature is likely to be interested in, or affected by, such dispute, the appropriate Government may, at the time of making the reference or at any time thereafter but before the submission of the award, include in that reference such establishment, group or class of establishments, whether or not at the time of such inclusion any dispute exists or is apprehended in that establishment, group or class of establishments.]

(6) Where any reference has been made under sub-section (1A) to a National Tribunal, then notwithstanding anything contained in this Act, no Labour Court or Tribunal shall have jurisdiction to adjudicate upon any matter which is under adjudication before the National Tribunal, and accordingly,--

- if the matter under adjudication before the National Tribunal is pending in a proceeding before a Labour Court or Tribunal, the proceeding before the Labour Court or the Tribunal, as the case may be, in so far as it relates to such matter, shall be deemed to have been quashed on such reference to the National Tribunal; and
- it shall not be lawful for the appropriate Government to refer the matter under adjudication before the National Tribunal to any Labour Court or Tribunal for adjudication during the pendency of the proceeding in relation to such matter before the National Tribunal.

(7) Where any industrial dispute, in relation to which the Central Government is not the appropriate Government, is referred to a National Tribunal, then notwithstanding anything contained in this Act, any reference in section 15, section 17, section 19, section 33A, section 33B and section 36A to the appropriate Government in relation to such dispute shall be construed as a reference to the Central Government but, save as aforesaid and as otherwise expressly provided in this Act, any reference in any other provision of this Act to the appropriate Government in relation to that dispute shall mean a reference to the State Government.

(8) No proceedings pending before a Labour Court, Tribunal or National Tribunal in relation to an industrial dispute shall lapse merely by reason of the death of any of the parties to the dispute being a workman, and such Labour Court, Tribunal or National Tribunal shall complete such proceedings and submit its award to the appropriate Government.

10A. Voluntary reference of disputes to arbitration.

(1). Where any industrial dispute exists or is apprehended and the employer and the workmen agree to refer the dispute to arbitration, they may, at any time before the dispute has been referred under section 10 to a Labour Court or Tribunal or National Tribunal, by a written agreement, refer the dispute to arbitration and the reference shall be to such person or persons (including the presiding officer of a Labour Court or Tribunal or National Tribunal) as an arbitrator or arbitrators as may be specified in the arbitration agreement.

(2) An arbitration agreement referred to in sub-section (1) shall be in such form and shall be signed by the parties thereto in such manner as may be prescribed.

(3) A copy of the arbitration agreement shall be forwarded to the appropriate Government and the conciliation officer and the appropriate Government shall, within one month from the date of the receipt of such copy, publish the same in the Official Gazette.

(4) The arbitrator or arbitrators shall investigate the dispute and submit to the appropriate Government the arbitration award signed by the arbitrator or all the arbitrators, as the case may be.

Where an industrial dispute has been referred to arbitration and a notification has been issued under sub-section (3A), the appropriate Government may, by order, prohibit the continuance of any strike or lock-out in connection with such dispute which may be in existence on the date of the reference.

(5) Nothing in the Arbitration Act, 1940 (10 of 1940), shall apply to arbitrations under this section.

CHAPTER IV

PROCEDURE, POWERS AND DUTIES OF AUTHORITIES

11. Procedure and powers of conciliation officers, Boards, Courts and Tribunals.-

(1) Subject to any rules that may be made in this behalf, an arbitrator, a Board, Court, Labour Court, Tribunal or National Tribunal shall follow such procedure as the arbitrator or other authority concerned may think fit.

(2) A conciliation officer or a member of a Board, [or Court or the presiding officer of a Labour Court, Tribunal or National Tribunal] may for the purpose of inquiry into any existing or apprehended industrial dispute, after giving reasonable notice, enter the premises occupied by any establishment to which the dispute relates.

(3) Every Board, Court, [Labour Court, Tribunal and National Tribunal] shall have the same powers as are vested in a Civil Court under the Code of Civil Procedure, 1908 (5 of 1908), when trying a suit, in respect of the following matters, namely:-

(a) enforcing the attendance of any person and examining him on oath;

(b) compelling the production of documents and material objects;

(c) issuing commissions for the examination of witnesses;

(d) in respect of such other matters as may be prescribed; and every inquiry or investigation by a Board, Court, [Labour Court, Tribunal or National Tribunal], shall be deemed to be a judicial proceeding within the meaning of sections 193 and 228 of the Indian Penal Code (45 of 1860).

(4) A conciliation officer may enforce the attendance of any person for the purpose of examination of such person or call for and inspect any document which he has ground for considering to be relevant to the industrial dispute [or to be necessary for the purpose of verifying the implementation of any award or carrying out any other duty imposed on him under this Act, and for the aforesaid

purposes, the conciliation officer shall have the same powers as are vested in a Civil Court under the Code of Civil Procedure, 1908 (5 of 1908), in respect of enforcing the attendance of any person and examining him or of compelling the production of documents.

(5) A Court, Labour Court, Tribunal or National Tribunal may, if it so thinks fit, appoint one or more persons having special knowledge of the matter under consideration as assessor or assessors to advise it in the proceeding before it.

(6) All conciliation officers, members of a Board or Court and the presiding officers of a Labour Court, Tribunal or National Tribunal shall be deemed to be public servants within the meaning of section 21 of the Indian Penal Code (45 of 1860).

(7) Subject to any rules made under this Act, the costs of, and incidental to, any proceeding before a Labour Court, Tribunal or National Tribunal shall be in the discretion of that Labour Court, Tribunal or National Tribunal and the Labour Court, Tribunal or National Tribunal, as the case may be, shall have full power to determine by and to whom and to what extent and subject to what conditions, if any, such costs are to be paid, and to give all necessary directions for the purposes aforesaid and such costs may, on application made to the appropriate Government by the person entitled, be recovered by that Government in the same manner as an arrear of land revenue.

(8) Every [Labour Court, Tribunal or National Tribunal] shall be deemed to be Civil Court for the purposes of [sections 345, 346 and 348 of the Code of Criminal Procedure, 1973 (5 of 1898).]

12. Duties of conciliation officers.-

(1) Where any industrial dispute exists or is apprehended, the conciliation officer may, or where the dispute relates to a public utility service and a notice under section 22 has been given, shall hold conciliation proceedings in the prescribed manner.

(2) The conciliation officer shall, for the purpose of bringing about a settlement of the dispute, without delay, investigate the dispute and all matters affecting the merits and the right settlement thereof and may do all such things as he thinks fit for the purpose of inducing the parties to come to a fair and amicable settlement of the dispute.

(3) If a settlement of the dispute or of any of the matters in dispute is arrived at in the course of the conciliation proceedings the conciliation officer shall send a report thereof to the appropriate Government [or an officer authorised in this behalf by the appropriate Government] together with a memorandum of the settlement signed by the parties to the dispute.

(4) If no such settlement is arrived at, the conciliation officer shall, as soon as practicable after the close of the investigation, send to the appropriate Government a full report setting forth the steps taken by him for ascertaining the facts and circumstances relating to the dispute and for bringing about a settlement thereof,

together with a full statement of such facts and circumstances, and the reasons on account of which, in his opinion, a settlement could not be arrived at.

(5) If, on a consideration of the report referred to in sub-section (4) the appropriate government is satisfied that there is a case for reference to a Board, (Labour court, Tribunal or National tribunal), it may make such reference. Where the appropriate Government does not make such a reference it shall record and communicate to the parties concerned its reasons there for.

(6) A report under this section shall be submitted within fourteen days of the commencement of the conciliation proceedings or within such shorter period as may be fixed by the appropriate Government:

13. Duties of Board.-

(1) Where a dispute has been referred to a Board under this Act, it shall be the duty of the Board to endeavour to bring about a settlement of the same and for this purpose the Board shall, in such manner as it thinks fit and without delay, investigate the dispute and all matters affecting the merits and the right settlement thereof and may do all such things as it thinks fit for the purpose of inducing the parties to come to a fair and amicable settlement of the dispute.

(2) If a settlement of the dispute or of any of the matters in dispute is arrived at in the course of the conciliation proceedings, the Board shall send a report thereof to the appropriate Government together with a memorandum of the settlement signed by the parties to the dispute.

(3) If no such settlement is arrived at, the Board shall, as soon as practicable after the close of the investigation, send to the appropriate

Government a full report setting forth the proceedings and steps taken by the Board for ascertaining the facts and circumstances relating to the dispute and for bringing about a settlement thereof, together with a full statement of such facts and circumstances, its findings thereon, the reasons on account of which, in its opinion, a settlement could not be arrived at and its recommendations for the determination of the dispute.

(4) If, on the receipt of a report under sub-section (3) in respect of a dispute relating to a public utility service, the appropriate Government does not make a reference to a [Labour Court, Tribunal or National Tribunal] under section 10, it shall record and communicate to the parties concerned its reasons there for.

(5) The Board shall submit its report under this section within two months of the date, [on which the dispute was referred to it] or within such shorter period as may be fixed by the appropriate Government:

Provided that the appropriate Government may from time to time extend the time for the submission of the report by such further periods not exceeding two months in the aggregate:

Provided further that the time for the submission of the report may be extended by such period as may be agreed on in writing by all the parties to the dispute.

14. Duties of Courts. - A Court shall inquire into the matters referred to it and report thereon to the appropriate Government ordinarily within a period of six months from the commencement of its inquiry.

15. Duties of Labour Courts, Tribunals and National Tribunals. -

Where an industrial dispute has been referred to a Labour Court, Tribunal or National Tribunal for adjudication, it shall hold its proceedings expeditiously and shall, [within the period specified in the order referring such industrial dispute or the further period extended under the second proviso to sub-section (2A) of section 10], submit its award to the appropriate Government.

16. Form of report or award.-

(1) The report of a Board or Court shall be in writing and shall be signed by all the members of the Board or Court, as the case may be:

Provided that nothing in this section shall be deemed to prevent any member of the Board or Court from recording any minute of dissent from a report or from any recommendation made therein.

(2) The award of a Labour Court or Tribunal or National Tribunal shall be in writing and shall be signed by its presiding officer.

17. Publication of reports and awards.-

(1) Every report of a Board or Court together with any minute of dissent recorded therewith, every arbitration award and every award of a Labour Court, Tribunal or National Tribunal shall, within a period of thirty days from the date of its receipt by the appropriate Government, be published in such manner as the appropriate Government thinks fit.

(2) Subject to the provisions of section 17A, the award published under sub-section (1) shall be final and shall not be called in question by any Court in any manner whatsoever.

17A. Commencement of the award.-

(1) An award (including an arbitration award) shall become enforceable on the expiry of thirty days from the date of its publication under section 17:

Provided that--

(a) if the appropriate Government is of opinion, in any case where the award has been given by a Labour Court or Tribunal in relation to an industrial dispute to which it is a party; or

(b) if the Central Government is of opinion, in any case where the award has been given by a National Tribunal, that it will be inexpedient on public grounds affecting national economy or social justice to give effect to the whole or any part of the award, the appropriate Government, or as the case may be, the Central Government may, by notification in the Official Gazette, declare that the award shall not become enforceable on the expiry of the said period of thirty days.

(2) Where any declaration has been made in relation to an award under the proviso to sub-section (1), the appropriate Government or the Central Government may, within ninety days from the date of publication of the award under section 17, make an order rejecting or modifying the award, and shall, on the first available opportunity, lay the award together with a copy of the order before the Legislature of the State, if the order has been made by a State Government, or before Parliament, if the order has been made by the Central Government.

(3) Where any award as rejected or modified by an order made under sub-section (2) is laid before the Legislature of a State or before Parliament, such award shall become enforceable on the expiry of fifteen days from the date on which it is so laid; and where no order under sub-section (2) is made in pursuance of a declaration under the proviso to sub-section (1), the award shall become enforceable on the expiry of the period of ninety days referred to in sub-section (2).

(4) Subject to the provisions of sub-section (1) and sub-section (3) regarding the enforceability of an award, the award shall come into operation with effect from such date as may be specified therein, but where no date is so specified, it shall come into operation on the date when the award becomes enforceable under sub-section (1) or subsection (3), as the case may be.

17B. Payment of full wages to workman pending proceedings in higher courts.-Where in any case, a Labour Court, Tribunal or National Tribunal by its award directs reinstatement of any workman and the employer prefers any proceedings against such award in a High Court or the Supreme Court, the employer shall be liable to pay such workman, during the period of pendency of such proceedings in the High Court or the Supreme Court, full wages last drawn by him, inclusive of any maintenance allowance admissible to him under any rule if the workman had not been employed in any establishment during such period and an affidavit by such workman had been filed to that effect in such Court:

Provided that where it is proved to the satisfaction of the High Court or the Supreme Court that such workman had been employed and had been receiving adequate remuneration during any such period or part thereof, the Court shall order that no wages shall be payable under this section for such period or part, as the case may be.

18. Persons on whom settlements and awards are binding.-

(1) A settlement arrived at by agreement between the employer and workman otherwise than in the course of conciliation proceeding shall be binding on the parties to the agreement.

(2) Subject to the provisions of sub-section (3), an arbitration award which has become enforceable shall be binding on the parties to the agreement who referred the dispute to arbitration.

(3) A settlement arrived at in the course of conciliation proceedings under this Act [or an arbitration award in a case where a notification has been issued under sub-section (3A) of section 10A] or [an award [of a Labour Court, Tribunal or National Tribunal] which has become enforceable] shall be binding on--

(a) all parties to the industrial dispute;

(b) all other parties summoned to appear in the proceedings as parties to the dispute, unless the Board, [arbitrator,] [Labour Court, Tribunal or National Tribuna!], as the case may be, records the opinion that they were so summoned without proper cause;

(c) where a party referred to in clause (a) or clause (b) is an employer, his heirs, successors or assigns in respect of the establishment to which the dispute relates;

(d) where a party referred to in clause (a) or clause (b) is composed of workmen, all persons who were employed in the establishment or part of the establishment, as the case may be, to which the dispute relates on the date of the dispute and all persons who subsequently become employed in that establishment or part.

19. Period of operation of settlements and awards.-

(1) A settlement shall come into operation on such date as is agreed upon by the parties to the dispute, and if no date is agreed upon, on the date on which the memorandum of the settlement is signed by the parties to the dispute.

(2) Such settlement shall be binding for such period as is agreed upon by the parties, and if no such period is agreed upon, for a period of six months [from the date on which the memorandum of settlement is signed by the parties to the dispute], and shall continue to be binding on the parties after the expiry of the period aforesaid, until the expiry of two months from the date on which a notice

in writing of an intention to terminate the settlement is given by one of the parties to the other party or parties to the settlement.

(3) An award shall, subject to the provisions of this section, remain in operation for a period of one year [from the date on which the award becomes enforceable under section 17A Provided that the appropriate Government may reduce the said period and fix such period as it thinks fit:

Provided further that the appropriate Government may, before the expiry of the said period, extend the period of operation by any period not exceeding one year at a time as it thinks fit so, however, that the total period of operation of any award does not exceed three years from the date on which it came into operation.

(4) Where the appropriate Government, whether of its own motion or on the application of any party bound by the award, considers that since the award was made, there has been a material change in the circumstances on which it was based, the appropriate Government may refer the award or a part of it [to a Labour Court, if the award was that of a Labour Court or to a Tribunal, if the award was that of a Tribunal or of a National Tribunal] for decision whether the period of operation should not, by reason of such change, be shortened and the decision of [Labour Court or the Tribunal, as the case may be] on such reference shall, 7 be final.

(5) Nothing contained in sub-section (3) shall apply to any award which by its nature, terms or other circumstances does not impose, after it has been given effect to, any continuing obligation on the parties bound by the award.

(6) Notwithstanding the expiry of the period of operation under sub-section (3), the award shall continue to be binding on the parties until a period of two months has elapsed from the date on which notice is given by any party bound by the award to the other party or parties intimating its intention to terminate the award.

(7) No notice given under sub-section (2) or sub-section (6) shall have effect, unless it is given by a party representing the majority of persons bound by the settlement or award, as the case may be.

20. Commencement and conclusion of proceedings.-

(1) A conciliation proceeding shall be deemed to have commenced on the date on which a notice of strike or lock-out under section 22 is received by the conciliation officer or on the date of the order referring the dispute to a Board, as the case may be.

(2) A conciliation proceeding shall be deemed to have concluded--

(a) where a settlement is arrived at, when a memorandum of the settlement is signed by the parties to the dispute;

(b) where no settlement is arrived at, when the report of the conciliation officer is received by the appropriate Government or when the report of the Board is published under section 17, as the case may be; or

(c) when a reference is made to a Court, [Labour Court, Tribunal or National Tribunal] under section 10 during the pendency of conciliation proceedings. (3) Proceedings [before an arbitrator under section 10A or before a Labour Court, Tribunal or National Tribunal] shall be deemed to have commenced on the date of the [reference of the dispute for arbitration or adjudication, as the case may be] and such proceedings shall be deemed to have concluded [on the date on which the award becomes enforceable under section 17A]

21. Certain matters to be kept confidential.-

There shall not be included in any report or award under this Act any information obtained by a conciliation officer, Board, Court, [Labour Court, Tribunal, National Tribunal or an arbitrator] in the course of any investigation or inquiry as to a trade union or as to any individual business (whether carried on by a person, firm or company) which is not available otherwise than through the evidence given before such officer, Board, Court, [Labour Court, Tribunal, National Tribunal or arbitrator], if the trade union, person, firm or company, in question has made a request in writing to the conciliation officer, Board, Court, [Labour Court, Tribunal, National Tribunal or arbitrator], as the case may be, that such information shall be treated as confidential; nor shall such conciliation officer or any individual member of the Board, [or Court or the presiding officer of the Labour Court, Tribunal or National Tribunal or the arbitrator] or any person present at or concerned in the proceedings disclose any such information without the consent in writing of the secretary of the trade union or the person, firm or company in question, as the case may be:

Provided that nothing contained in this section shall apply to a disclosure of any such information for the purposes of a prosecution under section 193 of the Indian Penal Code (45 of 1860).

CHAPTER V

STRIKES AND LOCK-OUTS

22. Prohibition of strikes and lock-outs.-

(1) No person employed in a public utility service shall go on strike in breach of contract--

(a) without giving to the employer notice of strike, as herein-after provided, within six weeks before striking; or

(b) within fourteen days of giving such notice; or

(c) before the expiry of the date of strike specified in any such notice as aforesaid; or

(d) during the pendency of any conciliation proceedings before a conciliation officer and seven days after the conclusion of such proceedings.

(2) No employer carrying on any public utility service shall lock-out any of his workmen--

(a) without giving them notice of lock-out as hereinafter provided, within six weeks before locking out; or

(b) within fourteen days of giving such notice; or

(c) before the expiry of the date of lock-out specified in any such notice as aforesaid; or

(d) during the pendency of any conciliation proceedings before a conciliation officer and seven days after the conclusion of such proceedings.

(3) The notice of lock-out or strike under this section shall not be necessary where there is already in existence a strike or, as the case may be, lock-out in the public utility service, but the employer shall send intimation of such lock-out or strike on the day on which it is declared, to such authority as may be specified by the appropriate Government either generally or for a particular area or for a particular class of public utility services.

(4) The notice of strike referred to in sub-section (1) shall be given by such number of persons to such person or persons and in such manner as may be prescribed.

(5) The notice of lock-out referred to in sub-section (2) shall be given in such manner as may be prescribed.

(6) If on any day an employer receives from any persons employed by him any such notices as are referred to in sub-section (1) or gives to any persons employed by him any such notices as are referred to in sub-section (2), he shall within five days thereof report to the appropriate Government or to such authority as that Government may prescribe the number of such notices received or given on that day.

23. General prohibition of strikes and lock-outs.- No workman who is employed in any industrial establishment shall go on strike in breach of contract and no employer of any such workman shall declare a lock-out--

(a) during the pendency of conciliation proceedings before a Board and seven days after the conclusion of such proceedings;

(b) during the pendency of proceedings before [a Labour Court, Tribunal or National Tribunal] and two months after the conclusion of such proceedings;

(c) during any period in which a settlement or award is in operation, in respect of any of the matters covered by the settlement or award.

24. Illegal strikes and lock-outs.-

(1) A strike or a lock-out shall be illegal if--

(a) it is commenced or declared in contravention of section 22 or section 23; or

(b) it is continued in contravention of an order made under sub-section (3) of section 10 [or sub-section (4A) of section 10A].

(2) Where a strike or lock-out in pursuance of an industrial dispute has already commenced and is in existence at the time of the reference of the dispute to a Board, [an arbitrator, a] [Labour Court, Tribunal or National Tribunal], the continuance of such strike or lock-out shall not be deemed to be illegal, provided that such strike or lock-out was not at its commencement in contravention of the provisions of this Act or the continuance thereof was not prohibited under sub-section (3) of section 10 [or sub-section (4A) of section 10A].

(3) A lock-out declared in consequence of an illegal strike or a strike declared in consequence of an illegal lock-out shall not be deemed to be illegal. Prohibition of financial aid to illegal strikes and lock-outs.

CHAPTER VA

LAY-OFF AND RETRENCHMENT

25. Prohibition of financial aid to illegal strikes and lockouts.- No person shall knowingly expend or apply any money in direct furtherance of support of any illegal strike or lock-out.

25A. Application of sections 25C to 25E.-

(1) Sections 25C to 25E inclusive [shall not apply to industrial establishments to which Chapter VB applies, or--]

(a) to industrial establishments in which less than fifty workmen on an average per working day have been employed in the preceding calendar month; or

(b) to industrial establishments which are of a seasonal character or in which work is performed only intermittently.

(2) If a question arises whether an industrial establishment is of a seasonal character or whether work is performed therein only intermittently, the decision of the appropriate Government thereon shall be final.

25B. Definition of continuous service.-

For the purposes of this Chapter,--

(1) a workman shall be said to be in continuous service for a period if he is, for that period, in uninterrupted service, including service which may be interrupted on account of sickness or authorised leave or an accident or a strike which is not illegal, or a lock-out or a cessation of work which is not due to any fault on the part of the workman;

(2) where a workman is not in continuous service within the meaning of clause (1) for a period of one year or six months, he shall be deemed to be in continuous service under an employer--

(a) for a period of one year, if the workman, during a period of twelve calendar months preceding the date with reference to which calculation is to be made, has actually worked under the employer for not less than--

(i) one hundred and ninety days in the case of a workman employed below ground in a mine; and

(ii) two hundred and forty days, in any other case;

(b) for a period of six months, if the workman, during a period of six calendar months preceding the date with reference to which calculation is to be made, has actually worked under the employer for not less than--

(i) ninety-five days, in the case of a workman employed below ground in a mine; and

(ii) one hundred and twenty days, in any other case.

Explanation.--For the purposes of clause (2), the number of days on which a workman has actually worked under an employer shall include the days on which--

(i) he has been laid-off under an agreement or as permitted by standing orders made under the Industrial Employment (Standing Orders) Act, 1946 (20 of 1946), or under this Act or under any other law applicable to the industrial establishment;

(ii) he has been on leave with full wages, earned in the previous years;

(iii) he has been absent due to temporary disablement caused by accident arising out of and in the course of his employment; and

(iv) in the case of a female, she has been on maternity leave; so, however, that the total period of such maternity leave does not exceed twelve weeks.]

25C. Right of workmen laid-off for compensation.-

Whenever a workman (other than a badli workman or a casual workman) whose name is borne on the muster rolls of an industrial establishment and who has completed not less than one year of continuous service under an employer is laid-off, whether continuously or intermittently, he shall be paid by the employer for all days during which he is so laid-off, except for such weekly holidays as may intervene, compensation which shall be equal to fifty per cent. of the total of the basic wages and dearness allowance that would have been payable to him had he not been so laid-off:

Provided that if during any period of twelve months, a workman is so laid-off for more than forty-five days, no such compensation shall be payable in respect of any period of the lay-off after the expiry of the first forty-five days, if there is an agreement to that effect between the workman and the employer:

Provided further that it shall be lawful for the employer in any case falling within the foregoing proviso to retrench the workman in accordance with the provisions contained in section 25F at any time after the expiry of the first forty-five days of the lay-off and when he does so, any compensation paid to the workman for having been laid off during the preceding twelve months may be set off against the compensation payable for retrenchment.

Explanation.--"Badli workman" means a workman who is employed in an industrial establishment in the place of another workman whose name is borne on the muster rolls of the establishment, but shall cease to be regarded as such for the purposes of this section, if he has completed one year of continuous service in the establishment.]

25D. Duty of an employer to maintain muster rolls of workmen.

Notwithstanding that workmen in any industrial establishment have been laid-off, it shall be the duty of every employer to maintain for the purposes of this Chapter a muster roll and to provide for the making of entries therein by workmen who may present themselves for work at the establishment at the appointed time during normal working hours.

25E. Workmen not entitled to compensation in certain cases.- No compensation shall be paid to a workman who has been laid-off--

(i) if he refuses to accept any alternative employment in the same establishment from which he has been laid-off, or in any other establishment belonging to the same employer situate in the same town or village or situate within a radius of five miles from the establishment to which he belongs, if, in the opinion of the employer, such alternative employment does not call for any special kill or previous experience and can be done by the workman, provided that the wages which would normally have been paid to the workman are offered for the alternative employment also;

(ii) if he does not present himself for work at the establishment at the appointed time during normal working hours at least once a day;

(iii) if such laying-off is due to a strike or slowing-down of production on the part of workmen in another part of the establishment.

25F. Conditions precedent to retrenchment of workmen.-

No workman employed in any industry who has been in continuous service for not less than one year under an employer shall be retrenched by that employer until--

(a) the workman has been given one month's notice in writing indicating the reasons for retrenchment and the period of notice has expired, or the workman has been paid in lieu of such notice, wages for the period of the notice:

(b) the workman has been paid, at the time of retrenchment, compensation which shall be equivalent to fifteen days' average pay [for every completed year of continuous service] or any part thereof in excess of six months; and

(c) notice in the prescribed manner is served on the appropriate Government [or such authority as may be specified by the appropriate Government by notification in the Official Gazette].

25FF. Compensation to workmen in case of transfer of undertakings.- Where the ownership or management of an undertaking is transferred, whether by agreement or by operation of law, from the employer in relation to that undertaking to a new employer, every workman who has been in continuous service for not less than one year in that undertaking immediately before such transfer shall be entitled to notice and compensation in accordance with the provisions of section 25F, as if the workman had been retrenched:

Provided that nothing in this section shall apply to a workman in any case where there has been a change of employers by reason of the transfer, if--

(a) the service of the workman has not been interrupted by such transfer;

(b) the terms and conditions of service applicable to the workman after such transfer are not in any way less favourable to the workman than those applicable to him immediately before the transfer; and

(c) the new employer is, under the terms of such transfer or otherwise, legally liable to pay to the workman, in the event of his retrenchment, compensation on the basis that his service has been continuous and has not been interrupted by the transfer.

Sixty days' notice to be given of intention to close down any undertaking.

25G. Procedure for retrenchment.-

Where any workman in an industrial establishment, who is a citizen of India, is to be retrenched and he belongs to a particular category of workmen in that establishment, in the absence of any agreement between the employer and the workman in this behalf, the employer shall ordinarily retrench the workman who was the last person to be employed in that category, unless for reasons to be recorded the employer retrenches any other workman.

25H. Re-employment of retrenched workmen.-

Where any workmen are retrenched, and the employer proposes to take into his employ any persons, he shall, in such manner as may be prescribed, give an opportunity [to the retrenched workmen who are citizens of India to offer themselves for re-employment and such retrenched workman] who offer themselves for re-employment shall have preference over other persons.

25J. Effect of laws inconsistent with this Chapter.-

(1) The provisions of this Chapter shall have effect notwithstanding anything inconsistent therewith contained in any other law [including standing orders made under the Industrial Employment (Standing Orders) Act, 1946 (20 of 1946)]:

[Provided that where under the provisions of any other Act or rules, orders or notifications issued there under or under any standing orders or under any award, contract of service or otherwise, a workman is entitled to benefits in respect of any matter which are more favourable to him than those to which he would be entitled under this Act, the workman shall continue to be entitled to the more favourable benefits in respect of that matter, notwithstanding that he receives benefits in respect of other matters under this Act.]

(2) For the removal of doubts, it is hereby declared that nothing contained in this Chapter shall be deemed to affect the provisions of any other law for the time being in force in any State in so far as that law provides for the settlement of industrial disputes, but the rights and liabilities of employers and workmen in so far as they relate to lay-off and retrenchment shall be determined in accordance with the provisions of this Chapter.

CHAPTER VB

SPECIAL PROVISIONS RELATING TO LAY-OFF, RETRENCHMENT AND CLOSURE

IN CERTAIN ESTABLISHMENTS

25K. Application of Chapter VB.-

(1) The provisions of this Chapter shall apply to an industrial establishment (not being an establishment of a seasonal character or in which work is performed

only intermittently) in which not less than [one hundred] workmen were employed on an average per working day for the preceding twelve months.

(2) If a question arises whether an industrial establishment is of a seasonal character or whether work is performed therein only intermittently, the decision of the appropriate Government thereon shall be final.

25L. Definitions.- For the purposes of this Chapter,--

(a) "industrial establishment" means--

 (i) a factory as defined in clause (m) of section 2 of the Factories Act, 1948 (63 of 1948);

 (ii) a mine as defined in clause (i) of subsection (1) of section 2 of the Mines Act, 1952 (35 of 1952); or

 (iii) a plantation as defined in clause (f) of section 2 of the Plantations Labour Act, 1951 (69 of 1951);

(b) no withstanding anything contained in sub-clause (ii) of clause (a) of section 2,--

 (i) in relation to any company in which not less than fifty-one per cent. of the paid-up share capital is held by the Central Government, or

 (ii) in relation to any corporation [not being a corporation referred to in sub-clause (i) of clause (a) of section 2] established by or under any law made by Parliament, the Central Government shall be appropriate Government.

25M. Prohibition of lay-off.-

(1) No workman (other than a badli workman or a casual workman) whose name is borne on the muster rolls of an industrial establishment to which this Chapter applies shall be laid-off by his employer except [with the prior permission of the appropriate Government or such authority as may be specified by that Government by notification in the Official Gazette (hereinafter in this section referred to as the specified authority), obtained on an application made in this behalf, unless such lay-off is due to shortage of power or to natural calamity, and in the case of a mine, such lay-off is due also to fire, flood, excess of inflammable gas or explosion].

(2) An application for permission under sub-section (1) shall be made by the employer in the prescribed manner stating clearly the reasons for the intended lay-off and a copy of such application shall also be served simultaneously on the workmen concerned in the prescribed manner.

(3) Where the workman (other than badli workmen or casual workmen) of an industrial establishment, being a mine, have been laidoff under sub-section (1) for reasons of fire, flood or excess of inflammable gas or explosion, the employer, in relation to such establishment, shall, within a period of thirty days

from the date of commencement of such lay-off, apply, in the prescribed manner, to the appropriate Government or the specified authority for permission to continue the lay-off.

(4) Where an application for permission under sub-section (1) or sub-section (3) has been made, the appropriate Government or the specified authority, after making such enquiry as it thinks fit and after giving a reasonable opportunity of being heard to the employer, the workmen concerned and the persons interested in such lay-off, may, having regard to the genuineness and adequacy of the reasons for such lay-off, the interests of the workmen and all other relevant factors, by order and for reasons to be recorded in writing, grant or refuse to grant such permission and a copy of such order shall be communicated to the employer and the workmen.

(5) Where an application for permission under sub-section (1) or sub-section (3) has been made and the appropriate Government or the specified authority does not communicate the order granting or refusing to grant permission to the employer within a period of sixty days from the date on which such application is made, the permission applied for shall be deemed to have been granted on the expiration of the said period of sixty days.

(6) An order of the appropriate Government or the specified authority granting or refusing to grant permission shall, subject to the provisions of sub-section (7), be final and binding on all the parties concerned and shall remain in force for one year from the date of such order.

(7) The appropriate Government or the specified authority may, either on its own motion or on the application made by the employer or any workman, review its order granting or refusing to grant permission under sub-section (4) or refer the matter or, as the case may be, cause it to be referred, to a Tribunal for adjudication: Provided that where a reference has been made to a Tribunal under this sub-section, it shall pass an award within a period of thirty days from the date of such reference.

(8) Where no application for permission under sub-section (1) is made, or where no application for permission under sub-section (3) is made within the period specified therein, or where the permission for any lay-off has been refused, such lay-off shall be deemed to be illegal from the date on which the workmen had been laid-off and the workmen shall be entitled to all the benefits under any law for the time being in force as if they had not been laid-off.

(9) Notwithstanding anything contained in the foregoing provisions of this section, the appropriate Government may, if it is satisfied that owing to such exceptional circumstances as accident in the establishment or death of the employer or the like, it is necessary so to do, by order, direct that the provisions of subsection

(1), or, as the case may be, sub-section (3) shall not apply in relation to such establishment for such period as may be specified in the order.]

25N. Conditions precedent to retrenchment of workmen.-

(1) No workman employed in any industrial establishment to which this Chapter applies, who has been in continuous service for not less than one year under an employer shall be retrenched by that employer until,--

(a) the workman has been given three months' notice in writing indicating the reasons for retrenchment and the period of notice has expired, or the workman has been paid in lieu of such notice, wages for the period of the notice; and

(b) the prior permission of the appropriate Government or such authority as may be specified by that Government by notification in the Official Gazette (hereafter in this section referred to as the specified authority) has been obtained on an application made in this behalf.

(2) An application for permission under sub-section (1) shall be made by the employer in the prescribed manner stating clearly the reasons for the intended retrenchment and a copy of such application shall also be served simultaneously on the workmen concerned in the prescribed manner.

(3) Where an application for permission under sub-section (1) has been made, the appropriate Government or the specified authority, after making such enquiry as it thinks fit and after giving a reasonable opportunity of being heard to the employer, the workmen concerned and the persons interested in such retrenchment, may, having regard to the genuineness and adequacy of the reasons stated by the employer, the interests of the workmen and all other relevant factors, by order and for reasons to be recorded in writing, grant or refuse to grant such permission and a copy of such order shall be communicated to the employer and the workmen.

(4) Where an application for permission has been made under subsection (1) and the appropriate Government or the specified authority does not communicate the order granting or refusing to grant permission to the employer within a period of sixty days from the date on which such application is made, the permission applied for shall be deemed to have been granted on the expiration of the said period of sixty days.

(5) An order of the appropriate Government or the specified authority granting or refusing to grant permission shall, subject to the provisions of sub-section (6), be final and binding on all the parties concerned and shall remain in force for one year from the date of such order.

(6) The appropriate Government or the specified authority may, either on its own motion or on the application made by the employer or any workman, review its

order granting or refusing to grant permission under sub-section (3) or refer the matter or, as the case may be, cause it to be referred, to a Tribunal for adjudication: Provided that where a reference has been made to a Tribunal under this sub-section, it shall pass an award within a period of thirty days from the date of such reference.

(7) Where no application for permission under sub-section (1) is made, or where the permission for any retrenchment has been refused, such retrenchment shall be deemed to be illegal from the date on which the notice of retrenchment was given to the workman and the workman shall be entitled to all the benefits under any law for the time being in force as if no notice had been given to him.

(8) Notwithstanding anything contained in the foregoing provisions of this section, the appropriate Government may, if it is satisfied that owing to such exceptional circumstances as accident in the establishment or death of the employer or the like, it is necessary so to do, by order, direct that the provisions of subsection (1) shall not apply in relation to such establishment for such period as may be specified in the order.

(9) Where permission for retrenchment has been granted under subsection (3) or where permission for retrenchment is deemed to be granted under sub-section (4), every workman who is employed in that establishment immediately before the date of application for permission under this section shall be entitled to receive, at the time of retrenchment, compensation which shall be equivalent to fifteen days' average pay for every completed year of continuous service or any part thereof in excess of six months.]

25-O. Procedure for closing down an undertaking.-

(1) An employer who intends to close down an undertaking of an industrial establishment to which this Chapter applies shall, in the prescribed manner, apply, for prior permission at least ninety days before the date on which the intended closure is to become effective, to the appropriate Government, stating clearly the reasons for the intended closure of the undertaking and a copy of such application shall also be served simultaneously on the representatives of the workmen in the prescribed manner:

Provided that nothing in this sub-section shall apply to an undertaking set up for the construction of buildings, bridges, roads, canals, dams or for other construction work.

(2) Where an application for permission has been made under subsection (1), the appropriate Government, after making such enquiry as it thinks fit and after giving a reasonable opportunity of being heard to the employer, the workmen and the persons interested in such closure may, having regard to the genuineness and adequacy of the reasons stated by the employer, the interests of the general public and all other relevant factors, by order and for reasons to

be recorded in writing, grant or refused to grant such permission and a copy of such order shall be communicated to the employer and the workman.

(3) Where an application has been made under sub-section (1) and the appropriate Government does not communicate the order granting or refusing to grant permission to the employer within a period of sixty days from the date on which such application is made, the permission applied for shall be deemed to have been granted on the expiration of the said period of sixty days.

(4) An order of the appropriate Government granting or refusing to grant permission shall, subject to the provisions of sub-section (5) be final and binding on all the parties and shall remain in force for one year from the date of such order.

(5) The appropriate Government may, either on its own motion or on the application made by the employer or any workman, review its order granting or refusing to grant permission under sub-section (2) or refer the matter to a Tribunal for adjudication: Provided that where a reference has been made to a Tribunal under this sub-section, it shall pass an award within a period of thirty days from the date of such reference.

(6) Where no application for permission under sub-section (1) is made within the period specified therein, or where the permission for closure has been refused, the closure of the undertaking shall be deemed to be illegal from the date of closure and the workmen shall be entitled to all the benefits under any law for the time being in force as if the undertaking had not been closed down.

(7) Notwithstanding anything contained in the foregoing provisions of this section, the appropriate Government may, if it is satisfied that owing to such exceptional circumstances as accident in the undertaking or death of the employer or the like it is necessary so to do, by order, direct that the provisions of sub-section (1) shall not apply in relation to such undertaking for such period as may be specified in the order.

(8) Where an undertaking is permitted to be closed down under sub-section (2) or where permission for closure is deemed to be granted under sub-section (3), every workman who is employed in that undertaking immediately before the date of application for permission under this section, shall be entitled to receive compensation which shall be equivalent to fifteen days' average pay for every completed year of continuous service or any part thereof in excess of six months.]

25P. Special provision as to restarting of undertakings closed down before commencement of the Industrial Disputes (Amendment) Act, 1976.-

If the appropriate Government is of opinion in respect of any undertaking of an industrial establishment to which this Chapter applies and which closed down before the commencement of the Industrial Disputes (Amendment) Act, 1976 (32 of 1976),--

(a) that such undertaking was closed down otherwise than on account of unavoidable circumstances beyond the control of the employer;

(b) that there are possibilities of restarting the undertaking;

(c) that it is necessary for the rehabilitation of the workmen employed in such undertaking before its closure or for the maintenance of supplies and services essential to the life of the community to restart the undertaking or both; and

(d) that the restarting of the undertaking will not result in hardship to the employer in relation to the undertaking, it may, after giving an opportunity to such employer and workmen, direct, by order published in the Official Gazette, that the undertaking shall be restarted within such time (not being less than one month from the date of the order) as may be specified in the order.

25Q. Penalty for lay-off and retrenchment without previous permission.-

Any employer who contravenes the provisions of section 25M or (1) of section 25N shall be punishable with imprisonment for a term which may extend to one month, or with fine which may extend to one thousand rupees, or with both.

25R. Penalty for closure.-

(1) Any employer who closes down an undertaking without complying with the provisions of sub-section (1) of section 25-O shall be punishable with imprisonment for a term which may extend to six months, or with fine which may extend to five thousand rupees, or with both.

(2) Any employer, who contravenes [an order refusing to grant permission to close down an undertaking under sub-section (2) of section 25-O or a direction given under section 25P], shall be punishable with imprisonment for a term which may extend to one year, or with fine which may extend to five thousand rupees, or with both, and where the contravention is a continuing one, with a further fine which may extend to two thousand rupees for every day during which the contravention continues after the conviction.

25S. Certain provisions of Chapter VA to apply to an industrial establishment to which this Chapter applies.-

The provisions of sections 25B, 25D, 25FF, 25G, 25H, and 25J in Chapter VA shall, so far as may be, apply also in relation to an industrial establishment to which the provisions of this Chapter apply.]

CHAPTER VC

UNFAIR LABOUR PRACTICES

25T. Prohibition of unfair labour practice.- No employer or workman or a trade union, whether registered under the Trader Unions Act, 1926 (16 of 1926), or not, shall commit any unfair labour practice.

25U. Penalty for committing unfair labour practices.-Any person who commits any unfair labour practice shall be punishable with imprisonment for a term which may extend to six months or with fine which may extend to one thousand rupees or with both.]

CHAPTER VI

PENALTIES

26. Penalty for illegal strikes and lock-outs.- (1) Any workman who commences, continues or otherwise acts in furtherance of, a strike which is illegal under this Act, shall be punishable with imprisonment for a term which may extend to one month, or with fine which may extend to fifty rupees, or with both.

(2) Any employer who commences, continues, or otherwise acts in furtherance of a lock-out which is illegal under this Act, shall be punishable with imprisonment for a term which may extend to one month, or with fine which may extend to one thousand rupees, or with both.

27. Penalty for instigation, etc: Any person who instigates or incites others to take part in, or otherwise acts in furtherance of, a strike or lock-out which is illegal under this Act, shall be punishable with imprisonment for a term which may extend to six months, or with fine which may extend to one thousand rupees, or with both.

28. Penalty for giving financial aid to illegal strikes and lock-outs.- Any person who knowingly expends or applies any money in direct furtherance or support of any illegal strike or lock-out shall be punishable with imprisonment for a term which may extend to six months, or with fine which may extend to one thousand rupees, or with both.

29. Penalty for breach of settlement or award.- Any person who commits a breach of any term of any settlement or award, which is binding on him under this Act, shall be punishable with imprisonment for a term which may extend to six months, or with fine, or with both, [and where the breach is a continuing one, with a further fine which may extend to two hundred rupees for every day during which the breach continues after the conviction for the first] and the Court trying the offence, if it fines the offender, may direct that the whole or any part of the fine realised from him shall be paid, by way of compensation, to any person who, in its opinion, has been injured by such breach.]

30. Penalty for disclosing confidential information.- Any person who willfully discloses any such information as is referred to in section 21 in contravention of the provisions of that section shall, on complaint made by or on behalf of the trade union or individual business affected, be punishable with imprisonment for a term which may extend to six months, or with fine which may extend to one thousand rupees, or with both.

31. Penalty for other offences.-

(1) Any employer who contravenes the provisions of section 33 shall be punishable with imprisonment for a term which may extend to six months, or with fine which may extend to one thousand rupees, or with both.

(2) Whoever contravenes any of the provisions of this Act or any rule made there under shall, if no other penalty is elsewhere provided by or under this Act for such contravention, be punishable with fine which may extend to one hundred rupees.

CHAPTER VII

MISCELLANEOUS

32. Offence by companies, etc.- Where a person committing an offence under this Act is a company, or other body corporate, or an association of persons (whether incorporated or not), every director, manager, secretary, agent or other officer or person concerned with the management thereof shall, unless he proves that the offence was committed without his knowledge or consent, be deemed to be guilty of such offence.

33. Conditions of service, etc., to remain unchanged under certain circumstances during pendency of proceedings.-

(1) During the pendency of any conciliation proceeding before a conciliation officer or a Board or of any proceeding before [an arbitrator or] a Labour Court or Tribunal or National Tribunal in respect of an industrial dispute, no employer shall-

(a) in regard to any matter connected with the dispute, alter, to the prejudice of the workmen concerned in such dispute, the conditions of service applicable to them immediately before the commencement of such proceeding; or

(b) for any misconduct connected with the dispute, discharge or punish, whether by dismissal or otherwise, any workmen concerned in such dispute, save with the express permission in writing of the authority before which the proceeding is pending.

(2) During the pendency of any such proceeding in respect of an industrial dispute, the employer may, in accordance with the standing orders applicable to a workman concerned in such dispute [or, where there are no such standing orders, in accordance with the terms of the contract, whether express or implied, between him and the workman-

(a) alter, in regard to any matter not connected with the dispute, the conditions of service applicable to that workman immediately before the commencement of such proceeding; or

(b) for any misconduct not connected with the dispute, or discharge or punish, whether by dismissal or otherwise, that workman:

Provided that no such workman shall be discharged or dismissed, unless he has been paid wages for one month and an application has been made by the employer to the authority before which the proceeding is pending for approval of the action taken by the employer.

(3) Notwithstanding anything contained in sub-section (2), no employer shall, during the pendency of any such proceeding in respect of an industrial dispute, take any action against any protected workman concerned in such dispute-

(a) by altering, to the prejudice of such protected workman, the conditions of service applicable to him immediately before the commencement of such proceedings; or

(b) by discharging or punishing, whether by dismissal or otherwise, such protected workman, save with the express permission in writing of the authority before which the proceeding is pending.

(4) In every establishment, the number of workmen to be recognised as protected workmen for the purposes of sub-section (3) shall be one per cent. of the total number of workmen employed therein subject to a minimum number of five protected workmen and a maximum number of one hundred protected workmen and for the aforesaid purpose, the appropriate Government may make rules providing for the distribution of such protected workmen among various trade unions, if any, connected with the establishment and the manner in which the workmen may be chosen and recognised as protected workmen.

(5) Where an employer makes an application to a conciliation officer, Board, [an arbitrator, a] labour Court, Tribunal or National Tribunal under the proviso to sub-section (2) for approval of the action taken by him, the authority concerned shall, without delay, hear such application and pass, [within a period of three months from the date of receipt of such application], such order in relation thereto as it deems fit:]

33A. Special provision for adjudication as to whether conditions of service, etc., changed during pendency of proceedings.-

Where an employer contravenes the provisions of section 33 during the pendency of proceedings [before a conciliation officer, Board, an arbitrator, a Labour Court, Tribunal or National Tribunal], any employee aggrieved by such contravention may, make a complaint in writing,

(a) to such conciliation officer or Board, and the conciliation officer or Board shall take such complaint into account in inediating in, and promoting the settlement of, such industrial dispute; and

(b) to such arbitrator, Labour Court, Tribunal or National Tribunal and on receipt of such complaint, the arbitrator, Labour Court, Tribunal or National Tribunal, as the case may be, shall adjudicate upon the complaint as if it were a dispute referred to or pending before it, in accordance with the provisions of this Act and shall submit his or its award to the appropriate Government and the provisions of this Act shall apply accordingly.]

33B. Power to transfer certain proceedings.-

(1) The appropriate Government may, by order in writing and for reasons to be stated therein, withdraw any proceeding under this Act pending before a Labour Court, Tribunal or National Tribunal and transfer the same to another Labour Court, Tribunal or National Tribunal, as the case may be, for the disposal of the proceeding and the Labour Court, Tribunal or National Tribunal to which the proceeding is so transferred may, subject to special directions in the order of transfer, proceed either de novo or from the stage at which it was so transferred: Provided that where a proceeding under section 33 or section 33A is pending before a Tribunal or National Tribunal, the proceeding may also be transferred to a Labour Court.

(2) Without prejudice to the provisions of sub-section (1), any Tribunal or National Tribunal, if so authorized by the appropriate Government, may transfer any proceeding under section 33 or section

(3) For the purposes of computing the money value of a benefit, the Labour Court may, if it so thinks fit, appoint a commissioner who shall, after taking such evidence as may be necessary, submit a report to the Labour Court and the Labour Court shall determine the amount after considering the report of the commissioner and other circumstances of the case.

(4) The decision of the Labour Court shall be forwarded by it to the appropriate Government and any amount found due by the Labour Court may be recovered in the manner provided for in sub-section (1).

(5) Where workmen employed under the same employer are entitled to receive from him any money or any benefit capable of being computed in terms of money, then, subject to such rules as may be made in this behalf, a single application for the recovery of the amount due may be made on behalf of or in respect of any number of such workmen. Explanation.--In this section "Labour Court" includes any court constituted under any law relating to investigation and settlement of industrial disputes in force in any State.]

34. Cognizance of offences.-

(1) No Court shall take cognizance of any offence punishable under this Act or of the abetment of any such offence, save on complaint made by or under the authority of the appropriate Government.

(2) No Court inferior to that of [a Metropolitan Magistrate or a Judicial Magistrate of the first class], shall try any offence punishable under this Act.

35. Protection of persons.-

(1) No person refusing to take part or to continue to take part in any strike or lock-out which is illegal under this Act shall, by reason of such refusal or by reason of any action taken by him under this section, be subject to expulsion from any trade union or society, or to any fine or penalty, or to deprivation of any right or benefit to which he or his legal representatives would otherwise be entitled, or be liable to be placed in any respect, either directly or indirectly, under any disability or at any disadvantage as compared with other members of the union or society, anything to the contrary in the rules of a trade union or society notwithstanding.

(2) Nothing in the rules of a trade union or society requiring the settlement of disputes in any manner shall apply to any proceeding for enforcing any right or exemption secured by this section, and in any such proceeding the Civil Court may, in lieu of ordering a person who has been expelled from membership of a trade union or society to be restored to membership, order that he be paid out of the funds of the trade union or society such sum by way of compensation or damages as that Court thinks just.

36. Representation of parties.-

(1) A workman who is a party to a dispute shall be entitled to be represented in any proceeding under this Act by--

(a) [any member of the executive or office bearer] of a registered trade union of which he is a member:

(b) [any member of the executive or other office bearer] of a federation of trade unions to which the trade union referred to in clause (a) is affiliated;

(c) where the worker is not a member of any trade union, by [any member of the executive or other office bearer] of any trade union connected with, or by any other workman employed in, the industry in which the worker is employed and authorized in such manner as may be prescribed.

(2) An employer who is a party to a dispute shall be entitled to be represented in any proceeding under this Act by--

(a) an officer of an association of employers of which he is a member;

(b) an officer of a federation of association of employers to which the association referred to in clause (a) is affiliated;

(c) where the employer is not a member of any association of employers, by an officer of any association of employers connected with, or by any other employer engaged in, the industry in which the employer is engaged and authorized in such manner as may be prescribed.

(3) No party to a dispute shall be entitled to be represented by a legal practitioner in any conciliation proceedings under this Act or in any proceedings before a Court.

(4) In any proceeding [before a Labour Court, Tribunal or National Tribunal], a party to a dispute may be represented by a legal practitioner with the consent of the other parties to the proceeding and [with the leave of the Labour Court, Tribunal or National Tribunal, as the case may be]

36A. Power to remove difficulties.-

(1) If, in the opinion of the appropriate Government any difficulty or doubt arises as to the interpretation of any provision of an award or settlement, it may refer the question to such Labour Court, Tribunal or National Tribunal as it may think fit.

(2) The Labour Court, Tribunal or National Tribunal to which such question is referred shall, after giving the parties an opportunity of being heard, decide such question and its decision shall be final and binding on all such parties.]

36B. Power to exempt.-

Where the appropriate Government is satisfied in relation to any industrial establishment or undertaking or any class of industrial establishments or undertakings carried on by a department of that Government that adequate provisions exist for the investigation and settlement of industrial disputes in respect of workmen employed in such establishment or undertaking or class of establishments or undertakings, it may, by notification in the Official Gazette, exempt, conditionally or unconditionally such establishment or undertaking or class of establishments or undertakings from all or any of the provisions of this Act.].

37. Protection of action taken under the Act.-

No suit, prosecution or other legal proceeding shall lie against any person for anything which is in good faith done or intended to be done in pursuance of this Act or any rules made there under.

38. Power to make rules.-

(1) The appropriate Government may, subject to the condition of previous publication, make rules for the purpose of giving effect to the provisions of this Act.

(2) In particular and without prejudice to the generality of the foregoing power, such rules may provide for all or any of the following matters, namely:--

(a) the powers and procedure of conciliation officers, Boards, Courts, [Labour Courts, Tribunals and National Tribunals] including rules as to the summoning of witnesses, the production of documents relevant to the subject-matter of an inquiry or investigation, the number of members necessary to form a quorum and the manner of submission of reports and awards;

(b) the constitution and functions of and the filling of vacancies in Works Committees, and the procedure to be followed by such Committees in the discharge of their duties;

(c) the allowances admissible to members of Courts [and Boards and presiding officers of Labour Courts, Tribunals and National Tribunals] and to assessors and witnesses;

(d) the ministerial establishment which may be allotted to a Court, Board, [Labour Court, Tribunal or National Tribunal] and the salaries and allowances payable to members of such establishments;

(e) the manner in which and the persons by and to whom notice of strike or lock-out may be given and the manner in which such notices shall be communicated;

(f) the conditions subject to which parties may be represented by legal practitioners in proceedings under this Act before a Court, [Labour Court, Tribunal or National Tribunal];

(g) any other matter which is to be or may be prescribed.

(3) Rules made under this section may provide that a contravention thereof shall be punishable with fine not exceeding fifty rupees.

House of Parliament while it is in session for a total period of thirty days which may be comprised in one session or in [two or more successive sessions, and if, before the expiry of the session immediately following the session or the successive sessions aforesaid] both Houses agree in making any modification in the rule, or both Houses agree that the rule should not be made, the rule shall thereafter have effect only in such modified form or be of no effect, as the case may be; so, however, that any such modification or annulment shall be without prejudice to the validity of anything previously done under that rule.]

39. Delegation of powers.-

The appropriate Government may, by notification in the Official Gazette, direct that any power exercisable by it under this Act or rules made there under shall, in relation to such matters and subject to such conditions, if any, as may be specified in the direction, be exercisable also,--

(a) where the appropriate Government is the Central Government, by such officer or authority subordinate to the Central Government or by the State Government or by such officer or authority subordinate to the State Government, as may be specified in the notification; and

(b) where the appropriate Government is a State Government, by such officer or authority subordinate to the State Government as may be specified in the notification.

40. Power to amend Schedules.-

(1) The appropriate Government may, if it is of opinion that it is expedient or necessary in the public interest so to do, by notification in the Official Gazette, add to the First Schedule any industry, and on any such notification being issued, the First Schedule shall be deemed to be amended accordingly.

(2) The Central Government may, by notification in the Official Gazette, add to or alter or amend the Second Schedule or the Third Schedule and on any such notification being issued, the Second Schedule or the Third Schedule, as the case may be, shall be deemed to be amended accordingly.

(3) Every such notification shall, as soon as possible after it is issued, be laid before the Legislature of the State, if the notification has been issued by a State Government, or before Parliament, if the notification has been issued by the Central Government.]

6.8 Code of Discipline in Industry

To maintain harmonious relations and promote industrial peace, a Code of Discipline has been laid down which applies to both public and private sector enterprises. It specifies various obligations for the management and the workers with the objective of promoting cooperation between their representatives.

The basic objectives of Code of Discipline are to:

- Maintain peace and order in industry.
- Promote constructive criticism at all levels of management and employment.
- Avoid work stoppage in industry
- Secure the settlement of disputes and grievances by a mutually agreed procedure
- Avoiding litigations
- Facilitate a free growth of trade unions
- Eliminate all forms of coercion, intimidation and violations of rules and regulations governing industrial relations.

The Code is based on the following principles:

- There should be no strike or lockout without prior notice.
- No unilateral action should be taken in connection with any industrial matter.
- Employees should follow go slow tactics
- No deliberate damage should be caused to a plant or property
- Acts of violations, intimidation and coercion should not be resorted
- The existing machinery for the settlement of disputes should be utilized.
- Actions that disturb cordial relationships should be avoided. To ensure better discipline in industry, management and unions agree on not indulging into various actions. These actions can b summarized as follows:

Management and Union(s) agree

- that no unilateral action should be taken in connection with any industrial matter and that should be settled at appropriate level
- that the existing machinery for settlement of disputes should be utilized with the utmost efficiency
- that there should be no strike or lock-out without prior notice
- that neither party will have recourse to coercion, intimidation, victimization or go –slow tactics
- that they will avoid litigation, sit-down and stay-in strikes and lock-outs
- that they will promote constructive co-operation between their representatives at all levels and as between workers themselves

- that they will establish upon a mutually agreed grievance procedure which will ensure a speedy and full investigation leading to settlement;
- that they will abide by various stages in the grievance procedure and take no arbitrary action which would by-pass this procedure; and

Management Agrees

- not to increase work-loads unless agreed upon or settled otherwise
- not to support or encourage any unfair labor practice such as discrimination and victimization of any employee
- to take prompt action for settlement of grievances and implementation of settlements, awards, decision and orders
- to take appropriate disciplinary action against its officers and members in cases where enquiries reveal that they were responsible for precipitate action by workers leading to indiscipline

Union agrees

- not to engage in any form of physical duress
- not to permit demonstrations which are not peaceful
- that their members will not engage or cause other employees to engage in any union activity during working hours
- to discourage unfair labor practices such as negligence of duty, damage to property and insubordination to take prompt action to implement awards, agreements, settlements and decisions

6.9 The Industrial employment (Standing order) Act, 1946

1. SHORT TITLE, EXTENT AND APPLICATION. -

(1) This Act may be called the Industrial Employment (Standing Orders) Act, 1946.

(2) It extends to the whole of India.

(3) It applies to every industrial establishment wherein one hundred or more workmen or employed, or were employed on any day of the preceding twelve months : Provided that the appropriate Government may, after giving not less than two months' notice of its intention so to do, by notification in the Official Gazette, apply the provisions of this Act to any industrial establishment employing such number of persons less than one hundred as may be specified in the notification :

(4) Nothing in this Act shall apply to -

(i) any industry to which the provisions of Chapter VII of the Bombay Industrial Relations Act, 1946 (11 of 1947), apply; or

(ii) any industrial establishment to which the provisions of the Madhya Pradesh Industrial Employment (Standing Orders) Act, 1961 (26 of 1961) apply:

Provided that notwithstanding anything contained in the Madhya Pradesh Industrial Employment (Standing Orders) Act, 1961 (26 of 1961), the provisions of this Act shall apply to all industrial establishments under the control of the Central Government.

2. INTERPRETATION. - In this Act, unless there is anything repugnant in the subject or context, -

(a) "appellate authority" means an authority appointed by the appropriate Government by notification in the Official Gazette to exercise in such area as may be specified in the notification the functions of an appellate authority under this Act : Provided that in relation to an appeal pending before an Industrial Court or other authority immediately before the commencement of the Industrial Employment (Standing Orders) Amendment Act, 1963 (39 of 1963), that Court or authority shall be deemed to be the appellate authority;

(b) "appropriate Government" means in respect of industrial establishments under the control of the Central Government or a Railway administration or in a major port, mine or oilfield, the Central Government, and in all other cases, the State Government :

Provided that where any question arises as to whether any industrial establishment is under the control of the Central Government, that Government may, either on a reference made to it by the employer or the workman or a trade union or other representative body of the workmen, or on its own motion and after giving the parties an opportunity of being heard, decide the question and such decision shall be final and binding on the parties;

(c) "Certifying Officer" means a Labour Commissioner or a Regional Labour Commissioner, and includes any other officer appointed by the appropriate Government, by notification in the Official Gazette, to perform all or any of the functions of a Certifying Officer under this Act;

(d) "employer" means the owner of an industrial establishment to which this Act for the time being applies, and includes -

(i) in a factory, any person named under clause (f) of sub-section (1) of section 7 of the Factories Act, 1948 (63 of 1948), as manager of the factory;

(ii) in any industrial establishment under the control of any department of any Government in India, the authority appointed by such Government in this behalf, or where no authority is so appointed, the head of the department;

(iii) in any other industrial establishment, any person responsible to the owner for the supervision and control of the industrial establishment;

(e) "industrial establishment" means -

(i) an industrial establishment as defined in clause (ii) of section 2 of the Payment of Wages Act, 1936 (4 of 1936), or

(ii) a factory as defined in clause (m) of section 2 of the Factories Act, 1948 (63 of 1948), or

(iii) a railway as defined in clause (4) of section 2 of the Indian Railways Act, 1890 (9 of 1890), or

(iv) the establishment of a person who, for the purpose of fulfilling a contract with the owner of any industrial establishment, employs workmen;

(f) "prescribed" means prescribed by rules made by the appropriate Government under this Act;

(g) "Standing orders" means rules relating to matters set out in the Schedule;

(h) "trade union" means a trade union for the time being registered under the Indian Trade Unions Act, 1926 (16 of 1926);

(i) "wages" and "workman" have the meanings respectively assigned to them in clauses (r) and (s) of section 2 of the Industrial Disputes Act, 1947 (14 of 1947).

3. SUBMISSION OF DRAFT STANDING ORDERS. -

(1) Within six months from the date on which this Act becomes applicable to an industrial establishment, the employer shall submit to the Certifying Officer five copies of the draft standing orders proposed by him for adoption in his industrial establishment.

(2) Provision shall be made in such draft for every matter set out in the Schedule which may be applicable to the industrial establishment, and where model standing orders have been prescribed, shall be, so far as is practicable, in conformity with such model.

(3) The draft standing orders submitted under this section shall be accompanied by a statement giving prescribed particulars of the workmen employed in the industrial establishment including the name of the trade union, if any, to which they belong.

(4) Subject to such conditions as may be prescribed, a group of employers in similar industrial establishments may submit a joint draft of standing orders under this section.

6. APPEALS. - (1) Any employer, workman, trade union or other prescribed representatives of the workmen aggrieved by the order of the Certifying Officer under sub-section (2) of section 5 may, within thirty days from the date on which copies are sent under sub-section (3) of that section, appeal to the appellate authority, and the appellate authority, whose decision shall be final, shall by order in writing confirm the standing orders either in the form certified by the Certifying Officer or after amending the said standing orders by making such modifications thereof or additions thereto as it thinks necessary to render the standing orders certifiable under this Act.

(2) The appellate authority shall, within seven days of its order under sub-section (1), send copies thereof of the Certifying Officer, to the employer and to the trade union or other prescribed representatives of the workmen, accompanied, unless it has confirmed without amendment the standing orders as certified by the Certifying Officer, by copies of the standing orders as certified by it and authenticated in the prescribed manner.

7. DATE OF OPERATION OF STANDING ORDERS. - Standing orders shall, unless an appeal is preferred under section 6, come into operation on the expiry of thirty days from the date on which authenticated copies thereof are sent under sub-section (3) of section 5, or where an appeal as aforesaid is preferred, on the expiry of seven days from the date on which copies of the order of the appellate authority, are sent under sub-section (2) of section 6.

8. REGISTER OF STANDING ORDERS. - A copy of all standing orders as finally certified under this Act shall be filed by the Certifying Officer in a register in the prescribed form maintained for the purpose, and the Certifying Officer shall furnish a copy thereof to any person applying therefore on payment of the prescribed fee.

9. POSTING OF STANDING ORDERS. - The text of the standing orders as finally certified under this Act shall be prominently posted by the employer in English and in the language understood by the majority of his workmen on special boards to be maintained for the purpose at or near the entrance through which the majority of the workmen enter the industrial establishment and in all departments thereof where the workmen are employed.

10. DURATION AND MODIFICATION OF STANDING ORDERS. - (1) Standing orders finally certified under this Act shall not, except on agreement between the employer and the workmen, or a trade union or other representative body of the workmen, be liable to modification until the expiry of six months from the date on which the standing orders or the last modifications thereof came into operation.

(2) Subject to the provisions of sub-section (1), an employer or workman or a trade union or other representative body of the workmen may apply to the Certifying Officer to have the standing orders modified, and such application shall be accompanied by five copies of the modifications proposed to be made, and where such modifications are proposed to be made by agreement between the employer and the workman or a trade union or other representative body of the workmen, a certified copy of that agreement shall be filed along with the application.

(3) The foregoing provisions of this Act shall apply in respect of an application under sub-section (2) as they apply to the certification of the first standing orders.

(4) Nothing contained in sub-section (2) shall apply to an industrial establishment in respect of which the appropriate Government is the Government of the State of Gujarat or the Government of the State of Maharashtra.

10A. PAYMENT OF SUBSISTENCE ALLOWANCE. - (1) Where any workman is suspended by the employer pending investigation or inquiry into complaints or charges of misconduct against him, the employer shall pay to such workman subsistence allowance - (a) at the rate of fifty per cent. of the wages which the workman was entitled to immediately preceding the date of such suspension, for the first ninety days of suspension; and

(b) at the rate of seventy-five percent of such wages for the remaining period of suspension if the delay in the completion of disciplinary proceedings against such workman is not directly attributable to the conduct of such workman.

(2) If any dispute arises regarding the subsistence allowance payable to a workman under sub-section (1), the workman or the employer concerned may refer the dispute to the Labour Court, constituted under the Industrial Disputes Act, 1947 (14 of 1947), within the local limits of whose jurisdiction the industrial establishment wherein such workman is employed is situate and the Labour Court to which the dispute is so referred shall, after giving the parties an opportunity of being heard, decide the dispute and such decision shall be final and binding on the parties.

(3) Notwithstanding anything contained in the foregoing provisions of this section, where provisions relating to payment of subsistence allowance under any other law for the time being in force in any State are more beneficial than the provisions of this section, the provisions of such other law shall be applicable to the payment of subsistence allowance in that State.

11. CERTIFYING OFFICERS AND APPELLATE AUTHORITIES TO HAVE POWERS OF CIVIL COURT. - (1) Every Certifying Officer and appellate authority shall have all the powers of a Civil Court for the purposes of receiving evidence, administering oaths, enforcing the attendance of witnesses, and compelling the discovery and production of documents, and shall be deemed to be a Civil Court within the meaning of sections 345 and 346 of the Code of Criminal Procedure, 1973 (2 of 1974).

(2) Clerical or arithmetical mistakes in any order passed by a Certifying Officer or appellate authority, or errors arising therein from any accidental slip or omission may, at any time, be corrected by that Officer or authority or the successor in office of such Officer or authority, as the case may be.

12. ORAL EVIDENCE IN CONTRADICTION OF STANDING ORDERS NOT ADMISSIBLE. - No oral evidence having the effect of adding to or otherwise varying or contradicting standing orders as finally certified under this Act shall be admitted in any Court.

12A. TEMPORARY APPLICATION OF MODEL STANDING ORDERS. - (1) Notwithstanding anything contained in sections 3 to 12, for the period commencing on the date on which this Act becomes applicable to an industrial establishment and ending with the date on which the standing orders as finally certified under this Act come into operation under section 7 in that establishment, the prescribed model standing orders shall be deemed to be adopted in that establishment, and the provisions of section 9, sub-section (2) of section 13 and section 13A shall apply to such model standing orders as they apply to the standing orders so certified.

(2) Nothing contained in sub-section (1) shall apply to an industrial establishment in respect of which the appropriate Government is the Government of the State of Gujarat or the Government of the State of Maharashtra.

13. PENALTIES AND PROCEDURE. - (1) An employer who fails to submit draft standing orders as required by section 3, or who modifies his standing orders otherwise than in accordance with section 10, shall be punishable with fine which may extend to five thousand rupees, and in the case of a continuing offence with a further fine which may extend to two hundred rupees for every day after the first during which the offence continues.

(2) An employer who does any act in contravention of the standing orders finally certified under this Act or his industrial establishment shall be punishable with fine which may extend to one hundred rupees, and in the case of a continuing offence with a further fine which may extend to twenty-five rupees for every day after the first during which the offence continues.

(3) No prosecution for an offence punishable under this section shall be instituted except with the previous sanction of the appropriate Government.

(4) No Court inferior to that of a Metropolitan Magistrate or Judicial Magistrate of the second class shall try any offence under this section.

13A. INTERPRETATION, ETC., OF STANDING ORDERS. - If any question arises as to the application or interpretation of a standing order certified under this Act, any employer or workman or a trade union or other representative body of the workmen may refer the question to any one of the Labour Courts constituted under the Industrial Disputes Act, 1947 (14 of 1947), and specified for the disposal of such proceeding by the appropriate Government by notification in the Official Gazette, and the Labour Court to which the question is so referred shall, after giving the parties an opportunity of being heard, decide the question and such decision shall be final and binding on the parties.

13B. ACT NOT TO APPLY TO CERTAIN INDUSTRIAL ESTABLISHMENTS. - Nothing in this Act shall apply to an industrial establishment in so far as the workmen employed therein are persons to whom the Fundamental and Supplementary Rules, Civil Services (Classification, Control and Appeal) Rules, Civil Services (Temporary Services) Rules, Revised Leave Rules, Civil Service Regulations, Civilians in Defence Service (Classification, Control and Appeal) Rules or the Indian Railway Establishment Code or any other rules or regulations that may be notified in this behalf by the appropriate Government in the Official Gazette, apply.

14. POWER TO EXEMPT. - The appropriate Government may by notification in the Official Gazette exempt, conditionally or unconditionally, any industrial establishment or class of industrial establishments from all or any of the provisions of this Act.

14A. DELEGATION OF POWERS. - The appropriate Government may, by notification in the Official Gazette, direct that any power exercisable by it under this Act or any rules made there under shall, in relation to such matters and subject to such conditions, if any, as may be specified in the direction, be exercisable also - (a) where the appropriate Government is the Central Government, by such officer or authority subordinate to the Central Government or by the State Government or by such officer or authority subordinate to the State Government, as may be specified in the notification;

(b) where the appropriate Government is a State Government, by such officer or authority subordinate to the State Government as may be specified in the notification.

15. POWER TO MAKE RULES. - (1) The appropriate Government may, after previous publication, by notification in the Official Gazette, make rules to carry out the purposes of this Act.

(2) In particular and without prejudice to the generality of the foregoing power, such rules may - (a) prescribe additional matters to be included in the Schedule, and the procedure to be followed in modifying standing orders certified under this Act in accordance with any such addition;

(b) set out model standing orders for the purposes of this Act;

(c) prescribe the procedure of Certifying Officers and appellate authorities;

(d) prescribe the fee which may be charged for copies of standing order entered in the register of standing orders;

(e) provide for any other matter which is to be or may be prescribed : Provided that before any rules are made under clause (a) representatives of both employers and workmen shall be consulted by the appropriate Government.

(3) Every rule made by the Central Government under this section shall be laid as soon as may be after it is made, before each House of Parliament while it is in session for a total period of thirty days which may be comprised in one session or in two or more successive sessions, and if, before the expiry of the session immediately following the session or the successive sessions aforesaid both Houses agree in making any modification in the rule or both Houses agree that the rule should not be made, the rule shall thereafter have effect only in such modified form or he of no effect, as the case may be; so however that any such modification or annulment shall be without prejudice to the validity of anything previously done under that rule.

6.10 Trade Unions (Amendment) Act, 2001

The Trade Unions Act, 1926 provides provisions for registration of trade unions of employers and workers and in certain respccts, it defines the law relating to registered trade unions. It confers legal and corporate status on registered trade unions. The Act is administered by the concerned State Governments.

The Trade Unions Act, 1926 has been amended and enforced w.e.f. 09.01.2002. The objective of these amendments, in brief, is to ensure orderly growth of trade unions and reduce multiplicity of trade unions and promote internal democracy.

6.11 Plantations Labour Act, 1951

Amendment proposals in the Plantations Labour Act, 1951 were discussed in the meeting of The Tripartite Industrial committee on Plantation Industry held on 03.04.2002 under the Chairmanship of Labour Minister. In pursuance of the decision taken in the meeting, an Inter-Ministerial committee with the representatives of the Ministries of Commerce, Finance and Labour and State Government of Assam, West Bengal, Kerala and Tamil Nadu was constituted on 05.06.2002 to look into the various issues concerning the plantation sector under the Chairmanship of Additional Secretary, Ministry of Commerce. The Inter-Ministerial Committee has since submitted the Report to the Government. The Report has been circulated to the concerned State Governments / Central Ministries / Departments. Amendment proposals in the Plantation Labour Act, 1951 are under consideration of the Government at present.

Summary

- An Industrial dispute is defined as a difference and disagreement in opinion between two disputants.
- According to section 2 (k) of Industrial Dispute Act, 1947, Industrial dispute is defined as."Any difference or disagreement between employers and employers or between employers and workmen, or between workmen and workmen, which is

connected with the employment or non-employment or the terms of employment or with conditions of Labour, of any person"

- There are two types of Industrial disputes – Interest disputes and Right disputes.
- Industrial disputes settlement involves following steps- collective bargaining, negotiation, conciliation and mediation, arbitration and adjudication.
- Collective bargaining is a technique by which disputes as to conditions of employment are resolved amicably by agreement rather than by coercion.
- Negotiation is one of the principle means of settling labour disputes.
- Strikes is defined as a cessation of work by a body of persons employed in any industry.
- A lockout is a temporary closing of a place of employment or the suspension of work or the refusal by an employer to continue to employ any number of persons employed by him.
- Types of strikes – General Strike, sympathetic strike, economic strike, stay in strike , slow down strikes, sick out –in , boycott , picketing , gherao.
- The Industrial disputes act was launched on 11th march 1947.
- The Industrial disputes act was an act to make provision for the investigation and settlement of industrial disputes and for certain other purposes.
- Code of discipline was established in Industry to maintain harmonious relations and promote industrial peace.

Objective Type Question

1. Difference and Disagreement in opinion between two disputants is called as

   (a) Grievances (b) Conflict

   (c) Industrial disputes (d) Strikes

2. Disputes related to determination of new wage level and other condition of employment-

   (a) Interest dispute (b) Rights disputes

   (c) Social disputes (d) Economical disputes

3. Industrial unrest manifest itself as-

   (a) Slow down strikes (b) Gherao

   (c) Boycott (d) All of above

4. Numbers of steps involved in collective bargaining-

   (a) one (b) Two

   (c) Three (d) Four

5. Cessation of work by a body of person employed in Industry-
   (a) Grievances (b) Conflict
   (c) Industrial disputes (d) Strikes
6. Temporary closing of a place of employment is called-
   (a) Strikes (b) Conflict
   (c) Lock outs (d) Grievances
7. Strike in which workers are dissuaded from work by certain workers who were standing at the gate of factory -
   (a) Economic strike (b) Slow down strike
   (c) Picketing (d) Boycott
8. Strike in which workers organises strike by remaining on their job-
   (a) Economic strike (b) Slow down strike
   (c) Picketing (d) Boycott
9. Industrial Disputes Act was formulated in which year?
   (a) 1947 (b) 1951
   (c) 1956 (d) 1948
10. The act of minimum wages was formulated in the year -
   (a) 1947 (b) 1951
   (c) 1956 (d) 1948
11. The industrial employment (standing orders) act was formulated in -
   (a) 1946 (b) 1947
   (c) 1956 (d) 1948
12. Compensation under the workmen's compensation act has to be paid -
   (a) with in one year (b) within one month
   (c) with in six month (d) Immediately

Review Questions

1. Define Industrial dispute.
2. Discuss the procedures for settling labour disputes.
3. Define strikes and mention its various classifications.
4. Write short notes on:
   (1) Sympathetic Strike
   (2) Economic strike
   (3) Gherao

5. Differentiate between slow down strike and sick out –in strike.
6. Enlist some salient features of the industrial disputes act 1947.
7. Define code of Disciplines in Industry with its objectives.
8. Define lock outs.
9. Give brief account on the trade union act 2001.
10. Give a brief account on the plantation labour act 1951.
11. Discuss the features of the industrial employment act 1946.

References

www.cottontown.org/page.cfm?pageid=603

www.franklinlakes.k12.nj.us/.../waterframe.html

http://www.uncp.edu/home/rwb/lecture_asian_civ.htm

www.wwnorton.com/.../ralph/resource/mughal.htm

http://www.indiadaily.org/images/green-revolution_60.jpg

http://www.sankalpindia.net/drupal/operation-flood-leading-white-revolution

http://www.jcbose.org/

http://gitasrujana.blogspot.com/2008/12/srinivasa-ramanujan.html

http://www.theindiapost.com/2009/08/12/dr-vikram-sarabhai-a-scientist-who-dared-to-dream/

http://www.thehindubusinessline.com/2004/06/10/images/2004061000051101.jpg

http://www.news.harvard.edu/gazette/2003/10.09/13-dunlop.html

http://www.teamtnt.com/images/grievanc.gif

http://www.ddws.nic.in/popups/SucessStory/Orrisa/sanjog/sanjog_index.html

http://www.visionmonday.com/CMSImagesContent/2008/7/workplace.jpg

http://www.historylearningsite.co.uk/domestic_system.htm

http://www.historyonthenet.com/Medieval_Life/feudalism.htm

http://www.makingthemodernworld.org.uk/stories/manufacture_by_machine/01.ST.01/

http://www2.sunysuffolk.edu/westn/hs-12_re.htm

http://industrialrevolution.sea.ca/causes.html

http://business.webindia.com/polinwid.htm

http://www.chillibreeze.com/articles/India-liberalization.asp

http://business.mapsofindia.com/fdi-india/investing-country/foreign-technology-collaboration.html

http://labour.delhigovt.nic.in/act/html_ida/introduction/introduction.html

http://labour.nic.in/annrep/annrep0405/english/Chapter3.pdf

http://www.legalserviceindia.com/articles/ind_dis.htm

http://laborsta.ilo.org/applv8/data/SSM7/E/IN.html

http://labour.nic.in/cgit/

http://industrialrelations.naukrihub.com/code-of-discipline.html

http://www.londonexternal.ac.uk/current_students/general_resources/general_docs/code_of_discipline.pdf

http://labour.delhigovt.nic.in/act/details_acts/industrial_employment/industrial_employ/chap_1.html

www.ignou.ac.in/edusat/mba/ms24/.../pdf/Unit-1.pdf

http://hrdjunction.blogspot.com/2009/06/approaches-of-industrial-relation.html

www.devenmeshram.blogspot.com

http://www.citehr.com/596-industrial-relation-india.html

http://www.ignou.ac.in/edusat/mba/ms24/Ms%2024%20Block%201/pdf/Unit-2.pdf

http://www.citehr.com/71361-future-strategies-industrial-relations-india.html

http://industrialrelations.naukrihub.com/measures-for-improving-industrial-relations.html

http://dipp.nic.in/evol1.htm

http://dcmsme.gov.in/policies/iip.htm#Indus1

http://siadipp.nic.in/publicat/ip/ipact01.htm

http://www.seiuhealthcarenj.org/for_members/What_is_a_Grievance_.aspx

http://www.openlearningworld.com/olw/courses/books/Employee%20Grievances%20and%20Discipline/Employee%20Grievances%20and%20Discipline/Model%20Grievance%20Procedure.html

http://www.openlearningworld.com/

http://www.management-hub.com/hr-grievance.html

http://managementhelp.org/intrpsnl/basics.htm

http://webhome.idirect.com/~kehamilt/ipsyconf.html

www.icbse.com

Britannica Online http://www.britannica.com/ Questia's The Columbia Encyclopedia http://www.questia.com/library/encyclopedia/ The Encyclopedia of World History http://www.bartleby.com/67/

The Internet History Sourcebooks Project

http://www.fordham.edu/halsall/ United Kingdom National Archives http://www.nationalarchives.gov.uk/

Washington State University, World Civilizations online class http://www.wsu.edu/~dee/

Wikipedia, the free encyclopedia http://en.wikipedia.org/wiki/Main_Page World History: 16th to 19th Centuries http://www.fsmitha.com/h3/index.html

B. Tech (Common to all branches)
Model Question Paper
Industrial Sociology

Time – 2 hours

Attempt all Questions

1. Attempt any three of the following:
   (a) Define the term Sociology. Who coined the term "Sociology"?
   (b) Discuss the relationship of Sociology with Psychology and Sociobiology.
   (c) How the phase of Industrialisation begins and highlights its effects in Asia.
   (d) What are the consequences of Industrial revolution?
2. Attempt any two of the following :
   (a) Discuss the value of Industrial Sociology in India.
   (b) What are the Productive systems? Explain any two of them.
   (c) Enlist the names of various inventors and inventions which lead to Industrial Revolution.
3. Attempt any two of the following:
   (a) What are the developments in Science after Independence in India?
   (b) Discuss the concept and scope of Industrial relations.
   (c) Discuss in brief the role of government in Industrial Relations.
4. Attempt any two of the following
   (a) Define Grievance, its causes and effects.
   (b) What are the types of conflict and managerial actions that causes workplace conflict?
   (c) Define any two of them
      (i) Industrial Disputes
      (ii) Strikes
      (iii) Lock outs

Index